Quantum Mechanics in Potential Representation and Applications

Quantum Mechanics in Potential Representation and Applications

Arvydas Juozapas Janavičius
Donatas Jurgaitis

Šiauliai University, Lithuania

 **World Scientific**

NEW JERSEY · LONDON · SINGAPORE · BEIJING · SHANGHAI · HONG KONG · TAIPEI · CHENNAI · TOKYO

Published by

World Scientific Publishing Co. Pte. Ltd.

5 Toh Tuck Link, Singapore 596224

USA office: 27 Warren Street, Suite 401-402, Hackensack, NJ 07601

UK office: 57 Shelton Street, Covent Garden, London WC2H 9HE

Library of Congress Control Number: 2020941539

British Library Cataloguing-in-Publication Data
A catalogue record for this book is available from the British Library.

QUANTUM MECHANICS IN POTENTIAL REPRESENTATION AND
APPLICATIONS

ISBN 978-981-121-665-7 (hardcover)
ISBN 978-981-121-666-4 (ebook for institutions)
ISBN 978-981-121-667-1 (ebook for individuals)

For any available supplementary material, please visit
https://www.worldscientific.com/worldscibooks/10.1142/11724#t=suppl

Typeset by Stallion Press
Email: enquiries@stallionpress.com

Preface

This book was written based on the new mathematical methods derived by Prof. (Hab. Dr.) A. J. Janavičius in 1977–1978 that were used for more exact optical model calculations of nuclei scattering in the Institute of Physics of Jagellonian University in Cracow. The method of potential representation that is developed allows us to transform the Schrödinger equation to the integral equation form and allows the wave function to be represented as the expression for a free or unperturbed solution multiplied on the multiplier depending on the perturbation potential. This method of simplified mathematical programs created for the optical model by Polish scientists Prof. K. Kwiatkowski and Prof. R. Planeta, and later while working in the Šiauliai University, Lithuania, by me and Prof. Dr. D. Jurgaitis, was used to find the relation between the scattering matrices for short-range forces and Coulomb potentials. Using the potential representation method, equations were obtained for a semi-relativistic single-particle shell model, and the stability and energy levels of the nucleons for the heaviest atomic nuclei $^{298}_{114}X$, $^{328}_{114}X$, $^{334}_{120}X$ and $^{340}_{126}X$ and also the energy levels of the charmed and bottom mesons in a semi-relativistic approach were evaluated.

We thank Monika Gruslytė for her help in improving the language. Authors' bibliographies are presented separately and are cited as [An].

Prof. (Hab. Dr.) A. J. Janavičius,
Prof. Dr. D. Jurgaitis

Introduction

The main author of the book, Prof. Hab. Dr. Arvydas Juozapas Janavičus is the founder of the new perturbation method in quantum mechanics, or potential representation method, commencing the research in this particular field in Krakow (Poland), Jagellonian University, and Institute of Nuclear Physics, during his scientific investigations in collaborating with Polish scientists Dr. K. Kwiatkowski and R. Planeta in 1977–1978. A. J. Janavičius derived a new scattering theory A[1, 2] (citations presented in Bibliography of the Authors) when wave functions and scattering matrices are defined directly by nucleon interaction potentials with nuclei. In 2000, Prof. A. J. Janavičius presented a series of lectures on the applications of potential representation method in nuclear physics while visiting the University of Modena and Reggio Emilia (Italy) in the framework of the Socrates–Erasmus academic exchange program. The second author, Prof. Hab. Dr. Donatas Jurgaitis, a scientist investigating differential equations, took part in a few investigations of the mathematically complicated tasks of scattering theory and semi-relativistic equation for nuclear physics. At the beginning of the investigations A[2], the wave function with orbital moment $l = 0$ for positive energies was presented as a product of the free solution and function, which depend on the potential of interaction. This allows one to find a rather handy analytical expression for scattering matrices for the Woods–Saxon potential. The obtained formula provides the analytical expression for the phase shifts that coincides with the calculations of the optical model. A general solution of the Schrödinger equation in the potential representation is obtained in the form of integral equations A[2] and A[3] for the positive and negative energies A[3]. The potential representation can be realized using the modified method of undefined coefficients. We found the integral equation A[4] for both separation of scattering matrices for short-radii potentials and the association between the scattering matrix and the concrete definite potential. The kernels of integral equations are Green's functions expressed by linearly independent solutions of Schrödinger equation for positive energies. Also, by applying this method, scattering matrices generated by Coulomb potentials can be separated from the scattering matrices for short-radius interaction potentials A[5]. Here, we have proved that the potential representation method is equivalent to the method of variation of constants for negative energies A[6]. The linearly independent solutions of the Schrödinger equation for harmonic oscillator potential have been obtained for the derivation of integral equations A[6], which are used for finding the eigenfunctions and eigenvalues for the Woods–Saxon potential for bound states. Moreover, the general solutions A[6] were obtained while multiplying the model wave function for the harmonic oscillator potential by the function which depends on the Woods–Saxon potential. The general solutions of the obtained integral equations [6] as model functions were used for calculating the neutron energy levels of ^{208}Pb nucleus for the Woods–Saxon potential. The integral equations were obtained for finding eigenfunctions and eigenvalues for spherical potentials A[7] using the method of indefinite coefficients A[8]. For bound states, wave functions can be expressed as a combination of the linearly independent unperturbed solutions for the model potential A[8] with the coefficients that depend on the additional interaction. We obtained Green's functions A[8] for solving the Schrödinger equation using the method of variation of constants. The linearly independent solutions using the harmonic oscillator model potential were obtained for the derivation of integral equations to find eigenfunctions and eigenvalues of the Schrödinger equation for the

Woods–Saxon potential A[9]. The kernels of the integral equations obtained are proportional to the difference between the perturbation and model potentials A[9].

Eigenvalues obtained by numerical iterations of these integral equations are in good agreement with the results that we obtained by the discretization method A[10]. The kernels of the obtained integral equations are proportional to the perturbation or difference of the Woods–Saxon and harmonic oscillator potentials A[9].

We considered wave functions in potential representation for hydrogen atoms using Hankel functions of the second kind A[11] as free solutions. The wave function in the potential representation for a helium atom was expressed as a product of both known solutions for the hydrogen atom and the function that depends on the interaction potential between electrons. A new form of equations in atomic spectroscopy was found. It was shown that when introducing the multipliers representing the interaction between electrons in He Hamiltonian, we obtained the solution and energy for the ground state of helium A[11].

In previous papers, the potential representation method of the perturbation theory for the discrete energies was proposed and considered. We applied this theory to the Jastrow's correlation method A[12]. It was demonstrated that by adding a new interparticle potential as perturbation to the Hamiltonian describing movement of these particles in the central field, we can obtain a new solution as a product of the unperturbed solution and the function which depends on this additional potential like the interaction between electrons e^2/r_{ik}. Using this method, we transformed the Schrödinger equation to the two related differential equations. First, we must solve the Schrödinger equation or use the known solution for a single particle in the central field. Then, we must solve a differential equation to find the function which depends on the interaction potential distances r_{ik} between electrons A[12]. Using this method, a new form of the Schrödinger equation for the atomic spectroscopy was found. This equation was also solved for the ground state of helium.

The presented multiplicative perturbation theory was successfully used in a semi-relativistic model of nuclei and in calculating its discrete energies. Usually, for calculating the energy levels of nuclei, the nonrelativistic Hartree–Fock equations with spin–orbit potential and effective interaction with the Woods–Saxon potential are used. One usually assumes that masses of nucleons in the shell model are constant, but this is not exact because mass of the nucleons depends on the state. Binding energy composes only 1% of the rest of the energy. If we calculate 1% from the estimated value of the kinetic energy, we will get a significant value. The relativistic energy corrections to mass for the harmonic oscillator potential A[13] of the ^{197}Au nucleus reached $-0.4\,\mathrm{MeV}$. The relativistic corrections to masses of nucleons and the Woods–Saxon potential A[14] were used to calculate the energies for all the states of the bound neutrons and protons of the ^{208}Pb nucleus. Here, a better correlation was obtained between one-nucleon binding energies and the energies obtained in the experiments. The calculated corrections for masses $-0.6788\,\mathrm{MeV}$ of neutrons and $-0.5884\,\mathrm{MeV}$ of protons for the energy levels of the ^{208}Pb nucleus for the excited states $1_{j15/2}$ and $1_{j13/2}$ are comparable with the energy of experimental levels. The relativistic corrections of potential energies are less than $0.05\,\mathrm{MeV}$ A[14]. The relativistic corrections for masses of nucleons are very important for the analysis of energy levels of nuclei of neutrons and protons, their scattering, stripping and knockout reactions. Relativistic corrections are important for calculating the scattering and the reaction parameters in the optical model and for the definition of parameters A[15] of mean potential in the nuclei. The new perturbation theory used for finding the solution of a semi-relativistic equation can significantly improve the calculations of the shell model and can be used for evaluating the stability of large nuclei A[16, 17] $^{298}_{114}X$, $^{328}_{114}X$, $^{334}_{120}X$ and $^{340}_{126}X$.

We present two different sets of physical solutions with different asymptotes at the origin for further investigations. A simple Hartree–Fock approach is presented for finding these solutions using the multiplicative or potential representation method of the perturbation theory A[18]. Important relativistic corrections for mass of nucleons are included. The presented semi-relativistic approach can be used in the consideration of heavy-nucleus energy levels and the stability of shells of a superheavy nucleus of protons and neutrons.

In previous papers, the relativistic corrections for the mass and potential energy to one-nucleon levels and the significant terms of the relativistic corrections for the mass of nucleons were obtained. In Ref. A[19], the exact mathematical method of a semi-relativistic model is considered. The semi-relativistic single-particle equation is a differential equation of the fourth order and it can be reduced to the integral–differential equations. The general solution of this equation must be expressed by the superposition of the four linearly independent solutions. Developing the modified Lagrange method and by using the multiplicative perturbation theory, we obtained the integral–differential equations for the wave functions with the usual asymptote at the origin r^{L+1} and unusual asymptote at r^{L+3}, r^{-L+2}. The wave functions with asymptote at the origin r^{L+3} must be used when the singular realistic nuclear potentials are included. The wave functions with different asymptotes give the different relativistic corrections for the mass and potential energy to one-nucleon levels. The general solution of this equation can be expressed by the superposition of the four linearly independent solutions. The additional asymptotes r^{-L+2}, r^{L+3} to r^{L+1} can be used for introducing new classification of quantum states for baryons A[20]. Instead of the classification using the three standard colors — red, blue, green — the states of quarks can be changed by the states having different asymptotes. For example, baryons $\Delta^{++}(uuu)$ and $\Delta^{-}(ddd)$ can be represented by the quarks displaced in states with asymptotes r^{L+1}, r^{-L+2}, r^{L+3} for $n = 1$, $L = 0$. Baryon $\Omega^{-}(sss)$ can be represented by the state r^{-L+2} for $n = 1$, $L = 1$, $m = -1, 0, +1$. There are no contradictions with the Pauli exclusion principle in the semi-relativistic case states.

Contents

Chapter 1

Quantum Nature of the Matter

1.1 The Structure of Atoms

Matter at the atomic and nuclear levels consists of particles, which have mass, charge, intrinsic angular momentum, or spin, and a magnetic moment associated with spin. All these parameters can take only discrete values. The stable electrons and nuclei combine to form atoms having discrete energy levels. Niels Bohr discovered that any observed discreteness required the introduction of the Planck's constant $\hbar = h/2\pi = 1.05457 \cdot 10^{-34}$ Js $= 6.58212 \cdot 10^{-22}$ Mevs. Atoms are the smallest part of elements that can exist. Atoms consist of a positive nuclei and electrons with mass $m_e = 9.1094 \cdot 10^{-31}$ kg that have a negative charge of $e = 1.6022 \cdot 10^{-19}$ C. Most materials are made up of neutral molecules formed by atoms held by chemical bonds with discrete energies and spatial displacements according to the combinations of spherical harmonics. This is the ground reason for the existence of genes, which can replicate and transmit biological information and are the reason for the existence of life. Atoms consist of negatively charged electrons that revolve around the nucleus with a density of about $2.3 \cdot 10^{17} \cdot$ kg/m^3 and a very small radius of $R = 1.25 \cdot A^{1/3}$ fm (Fermi length unit — fermi, $fm = 10^{-15}$ m), where A is the mass or nucleon number and is equal to the sum of neutral neutrons N and positive protons Z (charge number of the nucleus). The neutrons and protons have masses $m_n = 1.6749 \cdot 10^{-27}$ kg and $m_p = 1.6726 \cdot 10^{-27}$ kg, respectively, and are heavier by about 1840 times than the electrons, the mass of which is $m_e = 9.1094 \cdot 10^{-31}$ kg.

Now we can easily find the approximate density of the nucleus as follows:

$$\rho_n = \frac{{}^A_Z M}{V_n} \approx \frac{N \cdot m_n + Z \cdot m_p}{V_n},$$

$$V_n = \frac{4}{3} \cdot \pi \cdot R^3 = \frac{4}{3} \cdot \pi \cdot 1.25^3 \cdot A \cdot fm^3,$$

(1.1)

where the binding energy ${}^A_Z B$ is proportional to mass defect Δm, which indicates the difference in mass between the protons and neutrons, and the nucleus:

$${}^A_Z B = \Delta m \cdot c^2 = (N \cdot m_n + Z \cdot m_p) \cdot c^2 - {}^A_Z M c^2, \quad \Delta m > 0$$

(1.2)

which is less than the sum of the masses of free neutrons and protons. This difference in mass transforms to the binding energy of protons and neutrons in the nucleus and depends on A and Z. The binding energy ${}^A_Z B$ (1.2) and mass ${}^A_Z M$ of the nuclei can be calculated using the following semi-empirical formulas:

$${}^A_Z M = Z m_p + (A - Z) m_n - c_V A + c_S A^{2/3}$$
$$+ c_C Z^2 A^{-1/3} + c_A A^{-1}(A - 2Z) + \delta,$$

(1.3)

$${}^A_Z B = c_V A - c_S A^{2/3} - c_C Z^2 A^{-1/3} - c_A A^{-1}(A - 2Z) - \delta,$$

(1.4)

where the constants $c_V = 15.8$, $c_S = 18.0$, $c_C = 0.72$, $c_A = 23.5$, $\delta = c_P A^{-3/4}$ or 0, and $c_P = 33.5$ are expressed in MeV.

Binding energy (1.4), proportional to the number of nucleons $c_V A$, is created due to the interaction of the short-range force with a radius 1.4 fm. The second term $-c_S A^{2/3}$ represents the decrease in the binding energies of the nucleons on the surface of the nucleus. The third term defines the Coulomb repulsion energy. The term $-c_A A^{-1}(A-2Z)$ denotes the increasing stability of a nucleus that has an equal number of neutrons and protons. $\delta > 0$ for odd–odd nuclei, $\delta < 0$ for even–even nuclei, and $\delta = 0$ for odd–even nuclei. Using Equation (1.4), we estimate the binding energy of the ${}^{197}_{79}Au$ nucleus to be 1726.27 MeV. The binding energy for one nucleon is equal to 8.7628 MeV. The formula (1.4) cannot be applied to an atom with a small nucleus. From (1.2), we can find the defect of the nuclear mass as follows:

$$\Delta m = \frac{^{197}_{79}B}{c^2} = 3.0774 \cdot 10^{-27} \text{ kg},$$

$$eV = 1.6022 \cdot 10^{-19} \, J, \quad c = 2.99792 \cdot 10^8 \text{ m/s}.$$

(1.5)

Now from (1.2), we can find the mass of the $^{197}_{79}Au$ nucleus as follows:

$$^{197}_{79}M = (N \cdot m_n + Z \cdot m_p) - \Delta m = 326.70 \cdot 10^{-27} \text{ kg} \qquad (1.6)$$

and density for $A = 197$ is obtained as

$$\rho_n = \frac{^{199}_{79}M}{V_n} = 2.027 \cdot 10^{17} \text{ kg/m}^3. \qquad (1.7)$$

The overall binding energy in the nucleus is about 8.5 Mev and is about a million times greater than the binding energy of the electrons in atoms. Heavy nuclei such as $^{235}_{92}U$ and $^{239}_{94}Pu$ with an approximate binding energy per nucleon of 7.6 MeV have great fission energies, which are used for the production of atomic bombs and nuclear reactors.

For the $^{197}_{79}Au$ nucleus, we have $A = 197$, $Z = 79$ and $N = A - Z = 118$, and without including the mass defect (1.1), we obtain the density as

$$\rho_{Au} = 2.0471 \cdot 10^{17} \text{ kg/m}^3. \qquad (1.8)$$

The latter result is extraordinarily large if we compare it with the density of metallic gold, i.e., $19.3 \cdot 10^3 \cdot \text{kg/m}^3$. We can explain this difference by keeping in mind the fact that the radius of the $^{197}_{79}Au$ nucleus, which is calculated using the previously presented formula, is about $7.27 \cdot 10^{-15}$ m. It is very small compared with the radius $1.44 \cdot 10^{-10}$ m of the same atom. The difference of the binding energies of the electrons in atoms and nucleons in the nuclei is very significant. For example [2], the binding energies of neutrons and protons in the $^{197}_{79}Au$ nucleus in the external orbitals are 8.79 MeV (10^6 eV) and 7.73 MeV respectively. For the electrons in the external orbital [3], we have energy 9.23 eV for the same atom. We see that the binding energies ΔE_b of the nucleons in this case are about a million times greater than the binding energies of the electrons in atoms. For this reason, the chemical reaction energies are millions of times less than those for the nuclei.

1.2   The Schrödinger Equation

In quantum mechanics, the following Einstein's formulas for photons of energy E and momentum p

$$E = hv = \hbar\omega, \quad p = \hbar k, \quad k = \frac{2\pi}{\lambda}, \quad \lambda = h/p, \ \hbar = h/2\pi \quad (1.9)$$

were applied to waves of frequency v and for length λ by de Broglie. Schrödinger used de Broglie's idea that particles with momentum p are associated with a harmonic plane wave of length $\lambda = h/p$. Schrödinger applied it to the wave equation

$$\nabla^2\varphi(\vec{r}, t) - \frac{1}{u^2}\frac{\partial^2\varphi(\vec{r}, t)}{\partial t^2} = 0, \quad \varphi(\vec{r}, t) = e^{-i\omega t}\psi(\vec{r}) \quad (1.10)$$

for wave spreading with velocity $u = \lambda\omega/2\pi = \lambda v$. Then, the Schrödinger quantum equation for the wave function was obtained as follows:

$$\nabla^2\psi(\vec{r}) + \frac{4\pi^2}{\lambda^2}\psi(\vec{r}) = 0, \quad \frac{4\pi^2}{\lambda^2} = \frac{2m}{\hbar^2}[E - V(\vec{r})]. \quad (1.11)$$

The time-independent and time-dependent Schrödinger equations can be written in the following form:

$$-\frac{\hbar^2}{2m}\nabla^2\psi(\vec{r}) + V(\vec{r})\psi(\vec{r}) = E\psi(\vec{r}),$$

$$i\hbar\frac{\partial\psi(\vec{r}, t)}{\partial t} = -\frac{\hbar^2}{2m}\nabla^2\psi(\vec{r}, t) + V(\vec{r})\psi(\vec{r}, t). \quad (1.12)$$

The normalized wave function $\psi(\vec{r})$ in quantum mechanics according to Max Born's interpretation expresses the probability amplitude, absolute square of which defines the probability density of the particle as

$$\rho(\vec{r}) = \psi^*(\vec{r})\psi(\vec{r}), \int_V \psi^*(\vec{r})\psi(\vec{r})d^3\vec{r} = 1, \tag{1.13}$$

and for the bounded system, the mean values

$$\hat{L}^2 = \int_V \psi^*\hat{L}^2\psi d\bar{r}^3 \tag{1.14}$$

of any operator $\hat{L}$ commute with the Hamiltonian. For spherical symmetry, when potential $V(r) = V(\vec{r})$ depends only on the distance between two particles, we can obtain

$$[\hat{H}, \hat{L}^2] = \hat{H}\hat{L}^2 - \hat{L}^2\hat{H} = 0,$$

$$\hat{H} = \frac{\hat{p}^2}{2mr^2} + V(r) = -\frac{\hbar^2}{2mr^2}\frac{\partial}{\partial r}\left(r^2\frac{\partial}{\partial r}\right) + \frac{\hat{L}^2}{2mr^2} + V(r)$$

$$\tag{1.15}$$

and the following expression:

$$-\frac{\hbar^2}{2mr^2}\frac{\partial}{\partial r}\left(r^2\frac{\partial}{\partial r}\right)\psi + \frac{\vec{L}^2}{2mr^2}\psi + V(r)\psi = E\psi. \tag{1.16}$$

The central forces produce rotation around the origin. The operators of the total orbital angular momentum $\hat{L}$, impulse $\hat{p}$ and square $\hat{L}^2$ are related as follows:

$$\hat{L} = \vec{r} \times \hat{p} = \vec{r} \times \frac{\hbar}{i}\vec{\nabla}, \hat{L}^2 = r^2\vec{p}^2 + \hbar^2\frac{\partial}{\partial r}\left(r^2\frac{\partial}{\partial r}\right), \tag{1.17}$$

and in this case, they commute with the Hamiltonian and are conserved. They can be determined by the operators $L_Z, \hat{L}^2$, which are expressed in spherical coordinates as follows:

$$\hat{L}^2 = -\hbar^2\left[\frac{1}{\sin^2\vartheta}\frac{\partial^2}{\partial\varphi^2} + \frac{1}{\sin\vartheta}\frac{\partial}{\partial\vartheta}\left(\sin\vartheta\frac{\partial}{\partial\vartheta}\right)\right],$$

$$\hat{L}_z = \frac{\hbar}{i}\frac{\partial}{\partial\varphi}. \tag{1.18}$$

Eigenfunctions and eigenvalues of angular momentum operators were obtained by solving the equation

$$\hat{L}^2Y_l^m(\vartheta, \varphi) = \hbar^2\lambda_L Y_l^m(\vartheta, \varphi) \tag{1.19}$$

using the method of separation of variables as follows:

$$Y_l^m(\vartheta, \varphi) = \phi_m(\varphi)\theta_l^m(\vartheta), \quad \phi_m(\varphi) = e^{im\varphi}. \tag{1.20}$$

The obtained solution Y_l^m is normalized, with respect to an integration over the entire solid angle $\sin\vartheta d\vartheta d\varphi$. The orthonormal eigenfunctions, called spherical harmonics [4], are given as follows:

$$Y_l^m(\vartheta, \varphi) = \sqrt{\frac{2l+1}{4\pi}\frac{(l-m)!}{(l+m)!}} (-1)^m e^{im\varphi} P_l^m(\cos\vartheta),$$

$$\int_0^{2\pi}\int_0^\pi Y_l^m Y_{l1}^{m1} \sin\vartheta d\vartheta d\varphi = \delta_{l,l1}\delta_{m,m1},$$

$$\text{where } -l \leq m \leq l, \quad l = 0, 1, 2\ldots, \tag{1.21}$$

and satisfy the eigenvalues of $\hat{L}_Z$ and $\hat{L}^2$, i.e.,

$$\hat{L}_Z Y_l^m = \frac{\hbar}{i}\frac{\partial Y_l^m}{\partial\varphi} = m\hbar Y_l^m, \tag{1.22}$$

$$\hat{L}^2 Y_l^m = l(l+1)\hbar^2 Y_l^m. \tag{1.23}$$

Now inserting the wave function

$$\psi(\vec{r}) = \frac{u(r)}{r}Y_l^m(\vartheta, \varphi)$$

into (1.16), we obtain the radial Schrödinger equation

$$-\frac{\hbar^2}{2m}\frac{d^2}{dr^2}u(r) + \frac{\hbar^2}{2mr^2}u(r) + V(r)u(r) = Eu(r), \tag{1.24}$$

which can be simplified and presented as follows:

$$\hat{H}u(r) = Eu(r). \tag{1.25}$$

Keeping in mind Eq. (1.10), when the energy operator $\hat{E}$ is introduced to the fundamental equation for nonrelativistic quantum mechanics without spin can be justified as follows:

$$i\hbar\frac{\partial\psi}{\partial t} = -\frac{\hbar^2}{2m}\nabla^2\psi + V\psi, \quad \hat{E} = i\hbar\frac{\partial}{\partial t}. \tag{1.26}$$

The fact is that in quantum physics both waves and particles are described by the same wave equation (1.26). Hence, the position and momentum cannot be defined exactly simultaneously [4] and the following Heisenberg uncertainty principle is satisfied:

$$\Delta x \Delta p_x \geq \hbar. \tag{1.27}$$

If we represent a particle by a wave packet, we can obtain the last formula and arrive at the following second uncertainty principle [4]:

$$\Delta t = \frac{\Delta x}{v_x} \approx \frac{\hbar}{v_x \Delta p_x / 2} = \frac{\hbar}{\Delta E}, \tag{1.28}$$

$$\Delta E \cdot \Delta t \approx \hbar. \tag{1.29}$$

1.3 The Fundamental Forces

The defects of masses of nucleons constitute approximately about 1% of the rest of the masses and must be included in the expression of energy. For this, we will use the following expansion of relativistic energy [5]:

$$E = mc^2 = \frac{m_0 c^2}{\sqrt{1 - v^2/c^2}} \approx m_0 c^2 + \frac{1}{2} m_0 v^2 - \frac{1}{8} m_0 \frac{v^4}{c^2} + \cdots, \tag{1.30}$$

which can be used for a more exact expression of kinetic energy:

$$E - m_0 c^2 = \frac{1}{2} m_0 v^2 - \frac{1}{2} m_0 v^2 \cdot \frac{m_0 v^2}{4 m_0 c^2} + \cdots = \frac{p^2}{2m} - \frac{p^4}{8 m_0^3 c^2} + \cdots \tag{1.31}$$

The term $m_0 c^2$ represents the energy of the particle at rest. This energy in a nucleus has an enormous value. The rest of the mass represents a large accumulation of energy $m_0 = E/c^2$, which can be changed as $\Delta m_0 = \Delta E / c^2$ by changing of the energy ΔE. The relativistic improvement (1.31) for the kinetic energy operator

$$H_m = -\frac{p^4}{8 m_0^3 c^2}, \tag{1.32}$$

will be [3] important for calculating the average kinetic and bound energies of nucleons in the external orbitals of nuclei. The relativistic improvements for these energy levels of bound energies (1.31) of nucleons in the external shells significantly increase the binding

Table 1.1.   Fundamental forces.

Force	Strength	Theory	Mediator
Strong	10	Chromodynamics	Gluon
Electromagnetic	1/127	Electrodynamics	Photon
Weak	10^{-13}	Flavordynamics	W and Z
Gravitational	10^{-42}	Geometrodynamic	Graviton

energies and the stability of the superheavy nuclei A[16 and 17]. The relativistic correction to the kinetic energy or mass added to the Hamiltonian of the Schrödinger equation transforms it to a fourth-order differential equation whose exact solution needs the use of special mathematical methods [4].

There are just four fundamental forces in nature (Table 1.1): strong (nuclear), electromagnetic, weak and gravitational interactions where the three first act between the nuclei and the elementary particles. The interactions are classified according to the value of the characteristic dimensionless constants. The cross-section of the strong interaction is large and the interaction time is shorter. The strong interaction has a value of $R \approx 1.4 \cdot 10^{-13}$ cm = 1.4 fm. The one-pion exchange potential between nucleons has the following form:

$$V(r) = \frac{gh}{r} \exp\left(-\frac{r}{R_h}\right). \qquad (1.33)$$

Here, $R_h \approx \hbar/m_\pi c$ is the Compton wavelength of pion or distance of quantum diffusion in physical vacuum A[21, 22]. Then the dimensionless constant of interaction strength can be obtained from the formula

$$g_h^2/\hbar c \approx 1\_10. \qquad (1.34)$$

Strong interactions are realized by gluons, which are characterized by red, green and blue colors and their anti-colors. The interacting colored quarks change their colors or quantum states. This theory is called quantum chromodynamics. The weak interactions that cause changes in the flavors of the quarks are generated by vector bosons, of which two are charged (W^+ and W^-) and one is neutral (Z^0), that have great mass, i.e., 80.33 GeV and 91.19 GeV. For this case, the interaction radius is $R_w \approx \hbar/m_z c = 2.45 \cdot 10^{-16}$ cm, and the interaction strength is $g_w^2/\hbar c = 10^{-5}$ for a weak interaction and is described by a potential that is similar to that for strong interactions (3.4) with the constants alone differing.

For the electromagnetic interaction, the Coulomb potential is defined as

$$V(r) = \frac{1}{4\pi\varepsilon_0}\frac{e^2}{r} \tag{1.35}$$

for two particles having charge e, and the fine structure constant is given as

$$\alpha = e^2/\hbar c \approx 1/137. \tag{1.36}$$

Atoms consist of negative electrons and a positive, massive, very small nucleus, with diameter about 10 fm, which is placed in the center of the atom. J. J. Thompson and E. Rutherford found these particles, respectively in 1897 and 1911. The standard model of elementary particles asserts that the material in the universe is made up of quarks (Table 1.2) and leptons (Table 1.3). Quarks and leptons are fermions having spin $s = 1/2$. They interact by interchanging quants of these fields (Table 1.1). Bosons have spin $s = 1$. The fundamental forces, their strength and their interaction mediators [6] are listed in Table 1.1.

For investigating the atomic and molecular spectra that are generated by the electromagnetic forces based on the interchanging photons, quantum wave mechanics can be used. The quanta of the

Table 1.2.   Quark masses (MeV/c^2) and charges (e).

Quark flavor	Bare mass	Effective mass	Charges
d	5	340	$-1/3$
u	2	336	$2/3$
s	95	486	$-1/3$
c	1300	1550	$2/3$
b	4200	4730	$-1/3$
t	174000	177000	$2/3$

Table 1.3.   Leptons.

Leptons	Mass (MeV/c^2)	Mean life (s)	Electric charge (e)
Electron, e^-	0.5110	∞	$-e$
Electron neutrino, v_e	$<15 \cdot 10^{-6}$	∞?	0
Muon, μ^-	105.658	$2.197 \cdot 10^{-6}$	$-e$
Muon neutrino, v_μ	$0 < 0.17$	$\infty \cdot$?	0
Tau, τ^-	1777	$(291.0 \pm 1.5) \, 10^{-15}$	$-e$
Tau neutrino, μ_τ	<24	∞?	0

strong interaction field mediators in the hadrons and mesons are the gluons having zero mass and spin $s = 1$. The gluons explain the interaction that occurs between the three quarks in protons, neutrons and other more heavy baryons. Strong interacting mesons with protons and neutrons consist of quarks and antiquarks that interact by the interchange of gluons. The quanta of the strong interaction fields of gluons, unlike the photons, have rest masses equal to zero and are infinite force range and act on baryons and mesons that are confined by self-interactions similar to the quarks. Leptons interact through electromagnetic interactions if they are charged and interact with electrically neutral neutrinos through weak interactions. Quarks interact through strong, electromagnetic and weak interactions. The quanta of the weak interaction fields defining the interaction between leptons and quarks are the massive charged W^+, W^- and neutral Z bosons. According the Heisenberg uncertainty principle, a particle with mass m can exist in the intermediate state for a time $\hbar/mc^2$ and travel a distance no greater than $\hbar c/mc^2$. Since $m_W \approx 80 \cdot \text{GeV}/c^2$ and $m_Z \approx 90 \cdot \text{GeV}/c^2$, the weak interaction force has approximately a range of $10^{-3} \cdot$ fm. A lepton can change only into another lepton of the same type, i.e., $\mu^- \rightarrow v_\mu + e^- + \vec{v}_e$. In reactions with leptons, we have conservation lepton numbers that are positive for leptons and negative for antileptons. The leptons consist of negative electron e^- particles and positive positron antiparticles e^+ that are stable. From the assumption that neutrinos have zero masses, neutrinos can also be assumed to be stable. For the leptons, the reactions are valid only for the conservation of lepton numbers for antileptons, the reactions are considered to be negative. Also, whereas energies, charges, moments and spins must be conserved. The quarks, similar to the leptons, are Dirac fermions carrying charges $2e/3$ and $-e/3$. If quarks have a positive charge, then antiquarks have a negative charge. The number of quarks in an isolated system can never vary. The different types or flavors of quarks are not separately conserved because changes in quark flavors are possible only through weak interactions. Thus, with isolated quarks have never been observed in experiments. The existence of quarks was proved by considering the excited states of protons and neutrons, which appear as resonances in scattering experiments with photons and light mesons. The rich spectrum of the baryons can be described using a shell model of three confined quarks. Quarks are confined in compound systems in a region of space of about 1 fm. The most elementary quark systems are mesons consisting of a quark and an antiquark that have quark number 0. The protons, neutrons and other baryons have quark number 3. A proton having two *up* quarks with charge $2e/3$ and one *down* quark with charge $-e/3$ is stable. Neutrons have one *up* and two *down* quarks and have nonsignificantly more mass than protons, differing by about $1.3 \cdot \mathrm{MeV}/c^2$ in mass. In free space, neutrons with zero charge are unstable and decay to a proton through the weak interaction $n \rightarrow p + e^- + \bar{v}_e$ with a mean life of about $15 \cdot \mathrm{min}$. All mesons are unstable. The lightest mesons or pions are electrically charged $\pi^+ (u\bar{d})$, $\pi^- (\bar{u}d)$ and neutral π^0 consist of $u\bar{u}$ and $d\bar{d}$. The next lightest meson is η ($s\bar{s} \approx 547 \cdot \mathrm{MeV}$). For calculating the energy levels of electrons in an atom, Pauli's exclusion principle must be satisfied. This quantum mechanical principle must be applied to fermions, particles with spin $s = 1/2$ (intrinsic moment), requiring that any two electrons in an atom cannot be characterized by the same set of quantum numbers. The same principle must be applied to neutrons and protons in nuclei and for elementary quark particle [6] (Table 1.2), which compose the nucleons.

Quarks are point particles having spin $s = 1/2$ and similar to fermions must satisfy the Pauli's exclusion principle. But baryons Δ^{++} (uuu) in the lowest energy state with spins aligned in one direction $\uparrow\uparrow\uparrow$ are completely symmetric during the interchange of any two quarks, which is in contrast to the Pauli's principle. This problem was solved by introducing the indexes u, d, s, c, b and t for quark flavor in addition to three basic states of color indexes, named red, green and blue (r, g, b). We solve this problem A[20] by introducing into the Hamiltonian the relativistic correction for the mass of the bounded quarks. The asymptotic wave functions were usually obtained at the origin r^{L+1} and unusually at r^{-L+2}, r^{L+3}. For the ground state $J^P, \frac{3^+}{2}$ of hadrons Δ with quantum numbers $n = 1$ and $L = 0$, we can obtain three different eigenfunctions with three different asymptotes r^1, r^2 and r^3 when $r \to 0$ with different eigenvalues A[19, 20]. In addition to the first generation of light quarks $u(3\,\mathrm{MeV}/c^2)$, $d(7\,\mathrm{MeV}/c^2)$, we have the second generation $c(1200\,\mathrm{MeV}/c^2)$, $s(120\,\mathrm{MeV}/c^2)$ and the third generation $t(174000\,\mathrm{MeV}/c^2)$, $b(4300\,\mathrm{MeV}/c^2)$ with increasing masses and with the same charges $+\frac{2}{3}e$ and $-\frac{1}{3}e$. Hence, we have quarks of six different flavors (names). Baryons are all hadrons containing three quarks, and mesons contain quarks and antiquarks and have smaller masses. All these particles named hadrons take part in strong interactions. The quarks in hadrons interact by the exchange of gluons that have spin $s = 1$. Quarks are of different colors (red, green and blue) and interact with other quarks by interacting with bicolored (eight kinds) gluons. Thus, the colors and energies of quarks can be changed. These processes in hadrons can also be represented by eight transitions between the quantum states of wave functions [20] with asymptotes at the origin r^{L+1}, r^{-L+2}, r^{L+3} for orbital angular momentum quantum number values $L = 0, 1$. These quantum states of quarks with different definite energies inside the hadrons and mesons can lead to a formalistic change of the three-colors model that is more physically grounded. In the fundamental processes with interchanges with gluons $q \to q + g$, the colors of quarks may change but not the flavors. For example, a red quark can emit red–antiblue gluon, thus transforming itself into a blue quark and this gluon can transform a blue quark to red, thus realizing interchanging by a change of energies. In addition, the charged gluons can interact, thus producing gluon vertices and glueballs. The chromodynamic forces hold the quarks together, thus producing protons, neutrons and other baryons. The same strong forces form quarks and antiquarks that have mesons with a smaller mass. The constant of strong forces are larger than 1 and much more than constant for quantum electrodynamics $\alpha = \frac{e^2}{\hbar c} = \frac{1}{137}$. The nucleons in the nuclei interact mostly by changing the mesons because gluons like quarks have never been observed freely at distances greater than about 1 fermi. The light elementary particles named leptons (Table 1.3) do not take part in strong interactions.

The well-known leptons presented in Table 1.3 are fermions and have spin $s = 1/2$. Among the particles having rest masses, only electrons and positive positrons are stable. The muons μ^-, τ^- and the positive antiparticles differ from electrons and positrons only in their masses and their lifetimes. The difference in neutrino masses has not been clarified yet. e^-, μ^- and τ^- have been associated with different neutrinos V_e, V_μ and V_τ. In these interactions, the electrical charges are preserved and the leptons can change only with the same type of particles, i.e., $\mu^- \rightarrow v_\mu + e^- + \bar{v}_e$.

References

1. Jelley, N. A. (1990). *Fundamentals of Nuclear Physics*, Cambridge University Press, New York.
2. Shirokow, I. M. and Iudin, N. P. (1972). *Nuclear Physics*, Fizmatgiz, Moscow, p. 671 (in Russian).
3. Karazija, R. (1987). *The Theory of X-Ray and Electronic Spectra of Free Atoms. An Introduction*, Mokslas, Vilnius.
4. Merzbacher, E. (1970). *Quantum Mechanics*, John Wiley & Sons, New York.
5. Sitenko, A. G. and Tartakovskiy, V. N. (1972). *Lectures on Theory of Nucleus*, Atomizdat, Moscow, p. 351 (in Russian).
6. Griffiths, D. (2008). *Introduction to Elementary Particles*, Wiley-Vch Verlag Gmbh&Co. KgaA, Weinheim.

Chapter 2

Quantum Waves and Particles Diffusion in Physical Vacuum

2.1 Introduction

The new equation for quantum wave diffusion based on Heisenberg uncertainties and de Broglie frequency of waves is presented. Free movement and quantum diffusion through a rectangular barrier are considered. We find a quantum diffusion coefficient and radii of bound systems. The formula obtained connects the radii and bound energies of simple quantum systems, such as a hydrogen atom, deuteron and mesons, which consist of quarks.

Usually, in quantum mechanical investigations, properties of elementary particle systems are considered. Only the quantum field theory investigates the processes of interaction of real particles with virtual particles and antiparticles and thus represents the different fields in physical vacuum. Using quantum field interaction with particles, we can include the generation of particles and antiparticles and also the reactions that can be investigated experimentally. According to Sokolov and Tumanov [1], vacuum oscillations of the quantum field require the introduction of the effective radius of the electrons, which can help to explain Lamb shift of the atomic levels $2S_{1/2}$ and $2P_{1/2}$ [2] in hydrogen. The vacuum oscillations can scatter the dot-electron in a region with the radius R_e proportional to the

Compton wavelength [1] λ_e and the square root of the fine-structure constant α:

$$R_e = \sqrt{\alpha}\frac{\hbar}{m_0 c}, \quad \alpha = \frac{1}{4\pi\varepsilon_0}\frac{e^2}{\hbar c} = 7.29735257 \cdot 10^{-3},$$

$$\lambda_g = \frac{\hbar}{m_0 c} = 2.42 \cdot 10^{-10} \, \text{cm}.$$

(2.1)

The statistical model of quantum mechanics describes the movement of an electron in an atom that is similar to the movement of a Brownian particle A[21] interacting with fluctuations of electromagnetic vacuum. Considering that quantum phenomena have a stochastic character, we propose a new equation of quantum wave diffusion A[21] based on Heisenberg uncertainties and de Broglie waves [2]. The relation between uncertainties and non-locality [2] holds for all physical theories. Heisenberg observed that quantum mechanics [3] has the restrictions of accuracy of incompatible measurements such as position and momentum whose results cannot be predicted simultaneously. These restrictions are known as uncertainty relations. Applications of these uncertainties mainly to measurements are misleading because they suggest that the restrictions occur only when one makes measurements, but in our case, it is not necessary. Taking into consideration the Lamb shift and Eq. (2.1), we can say that the problem is more general than that in quantum mechanics. The definition of the wave–particle duality and the physical parameters requires modifying the classical Schrödinger equation based on the wave equation and de Broglie waves. We have proposed the quantum equation relating the stochastic quantum diffusion in physical vacuum and the de Broglie waves representing a guiding field for the direction of moving quantum particles.

2.2   Diffusion of Quantum Waves

We assume that the equation for quantum mechanics diffusion can be derived from the diffusion equations [4, A21], i.e.,

$$\frac{\partial \psi_J}{\partial t} = D_C \frac{\partial^2 \psi_J}{\partial x^2}, \quad D_C = \frac{\hbar}{4m} = \frac{\omega}{2k^2},$$

(2.2)

applied to the wave function

$$\psi_J = Ae^{-i\omega t + \lambda x}. \tag{2.3}$$

In this case, we obtain

$$\lambda^2 = -\frac{i\omega}{D_C}, \quad \lambda_{1,2} = \pm i\sqrt{\frac{\omega}{2D_C}} \mp \sqrt{\frac{\omega}{2D_C}}. \tag{2.4}$$

Requiring that the solution must represent some kind of linearly independent physical ψ_{J1} and nonphysical ψ_{J2} solutions, we have

$$\psi_{J1} = Ae^{-i\omega t + ikx - k|\Delta x|}, \quad \psi_{J2} = Be^{-i\omega t - ikx + k|\Delta x|}, \quad \Delta x \geq x - x_n,$$

$$k = \sqrt{\frac{\omega}{2D_C}}, \quad x_n = n\frac{\lambda}{2} \quad n = 0, 1, 2. \tag{2.5}$$

Free solutions can be obtained by considering the maximum ($x = x_0$) or minimum ($|x - x_0| = \pi/\Delta k$) amplitude for a wave packet or the following superposition of quantum oscillations (2.5):

$$\psi_J(x) = Ae^{ikx - k|x - x_0|} + Be^{-ikx + k|x - x_0|}, \tag{2.6}$$

where we can take $x = x_0$ for the coordinates at the maximum $x_{n0\,\text{max}}$ or minimum $x_{n0\,\text{min}}$ oscillations,

$$kx_{n0\,\text{max}} = n\pi, \quad n = 0, 1, 2, \ldots, |x - x_{n0\,\text{max}}| = 0, \tag{2.7}$$

$$kx_{n0\,\text{min}} = (2n + 1)\frac{\pi}{2}, \quad n = 0, 1, 2, \ldots, k|x_{n0\,\text{min}} - x_{n0\,\text{max}}| = \frac{\pi}{2}, \tag{2.8}$$

for the real parts of the wave function (2.6). We can represent this wave function at the point x by decreasing the oscillations generated at the maximum point $x_{n0\,\text{max}}$. Here, we have a train of decreasing waves similar to that seen in a wave packet.

Now we will try to consider the spreading wave in the x direction, i.e.,

$$\psi_J(x, t) = Ae^{-i\omega t + ikx - k|x - x_0|} = A\exp\left[\frac{i}{\hbar}(-Et + px + ip|x - x_0|)\right]. \tag{2.9}$$

This plain wave can be rewritten in the form:

$$\psi_J(x_0, t) = Ae^{-i(\omega \cdot t - kx_0) - k|x_0 - x|}, \tag{2.10}$$

which satisfies the simple wave equation

$$\psi_J(x_0, t) = Ae^{-i(\omega \cdot t - kx_0)}, \tag{2.11}$$

when $x = x_0$. We can represent this wave function at the point x_0 by spreading the oscillations generated at the maximum point (2.7) $x_{n0\,max}$, where we can find a moving particle with a maximum probability at the point x_0. Here, we have a train of decreasing waves similar to that seen in a wave packet where the proposed wave function (2.10) can be normalized by integrating the probability density $\psi^* \psi$

$$dP = A^2 e^{-2k|x-x_0|} d|x - x_0|,$$

$$P = 2A^2 \int_0^\infty e^{-2ky} dy = A^2 \frac{1}{k} = 1, \quad A = \sqrt{k}. \tag{2.12}$$

The probability of a freely moving particle to exist in the interval $d|x - x_0|$ is proportional to $k = \frac{2\pi}{\lambda}$, which is an important result of the scattering theory in quantum mechanics [2]. If a low-energy beam of particles is incident on a sphere with radius $r_0 = d|x - x_0|$, then from (2.12) we obtain $k \cdot r_0 < 1$ or $\hbar k \cdot r_0 < \hbar$. Only partial waves with orbital quantum numbers $l = 0$ take part in the interaction with a sphere, and freely moving particles represented by the wave function (2.10) are also located in this region. The velocity of these spreading waves can be evaluated but requiring

$$-Et + px + ip|x - x_0| = -Et_1 + px_1 + ip|x_1 - x_0|, \tag{2.13}$$

for equally complex phases when

$$t_1 = t_0 + \Delta_t, \quad x_1 = x_0 + \Delta x. \tag{2.14}$$

Substituting (2.14) in (2.13), we obtain

$$E\Delta t - p\Delta x - ip|\Delta x| = 0. \tag{2.15}$$

For the movement of the wave packet maximum $\Delta x_m > 0$ connected with particles generated by quantum diffusion in physical vacuum at the wave maximum point x_0, we have $\Delta x_m = \frac{1}{2}\Delta x = \frac{1}{2}|\Delta x|$.

Then, from the last equation for a non-relativistic case $E = p^2/2m$, we obtain

$$v_J = \frac{\Delta x_m}{\Delta t} = \frac{2E}{p + ip} = \frac{v}{2}(1 - i). \tag{2.16}$$

Calculating the square of modulus, $v_j{}^* v_J$, we note that for a free-spreading quantum diffusion stochastic wave train, the equation

$$v_J{}^* v_J = \frac{v^2}{2}, \tag{2.17}$$

satisfies the conservation of kinetic energy

$$m v_J{}^* v_J = \frac{m v^2}{2}, \tag{2.18}$$

and momentum mv for a freely moving quantum particle with the average velocity v and mass m. From this we can draw an assumption that

$$vt \approx x_{n0\,\mathrm{max}} = n\frac{\lambda}{2}, |\Delta x_n| = |x - x_{n0\,\mathrm{max}}| = \left| x - n\frac{\lambda}{2} \right|. \tag{2.19}$$

Finding the minimum difference $\Delta x_n = vt - n\frac{\lambda}{2} = x - n\frac{\lambda}{2} < \frac{\lambda}{2}$ from (2.16), (2.8) we can determine n, $x_{n0\,\mathrm{max}}$ and $|\Delta x_n|$. Also, the free solutions (2.9) and (2.7) of the quantum diffusion Eq. (2.2) are defined.

From this, for a free space [2] when $\omega = ck$ and $E = \hbar\omega$, we can obtain the wave function

$$\psi_{J1} = A \exp\left[-i\omega \cdot t + \frac{1}{c\hbar}(iEx - E|\Delta x_n|) \right], \tag{2.20}$$

which includes oscillations in physical vacuum. The free particle with mass m moves with a velocity v by the action of classical forces and quantum forces [6] depending on the wave function or in our case on the stochastic wave packet generated in physical vacuum at the point $n\frac{\lambda}{2}$ whose maximum amplitude is distributed with velocity v.

Comparing the standard formula $E = \frac{k^2\hbar^2}{2m}$ and (2.5), we have

$$k^2 = \frac{2mE}{\hbar^2} = \frac{\omega}{2D_C}, \quad E = \hbar\omega. \tag{2.21}$$

We obtain the quantum stochastic diffusion equation for a wave A[21]

$$\frac{\partial \psi_J}{\partial t} = D_C \frac{\partial^2 \psi_J}{\partial x^2}, \quad D_C = \frac{\hbar}{4m} = \frac{\omega}{2k^2}. \tag{2.22}$$

For free moving particles with $k^2 = \frac{2mE}{\hbar^2}$, from the last formula, we obtain the standard non-relativistic expression

$$E = \hbar\omega = \frac{p^2}{2m}, \quad p = \hbar k. \tag{2.23}$$

From the expression of the quantum diffusion coefficient, we can infer that a photon is reducible to virtual particles and antiparticles [6]. When an important expression (2.5) is satisfied, a connection with relativistic virtual processes [6] in physical vacuum

$$m = \frac{\hbar k^2}{2\omega} = \frac{h\nu}{2c^2}, \quad h\nu = 2mc^2 \tag{2.24}$$

can be obtained. The last equation shows that a photon can produce both a particle and an antiparticle with common mass $2m$ or annihilate by virtual processes.

We can also obtain the expression of the diffusion coefficient D_C from the Heisenberg uncertainties for oscillations in physical vacuum A[21] as follows:

$$2mc \cdot \Delta r = \hbar, \quad 2mc^2 \cdot \Delta t = \hbar, \tag{2.25}$$

$$D_C = \frac{1}{2}\Delta r^2 \frac{1}{\Delta t} = \frac{\hbar}{4m}, \quad \Delta t = D_C \frac{2}{c^2}. \tag{2.26}$$

From (2.24), we get

$$\omega \psi_{J1} = \frac{\hbar k^2}{2m} \psi_{J1}. \tag{2.27}$$

Multiplying the last equation by $\hbar$, if the de Broglie equation $\lambda = h/p$ is satisfied, we obtain

$$E \psi_{J1} = \frac{p^2}{2m} \psi_{J1}, \quad p = \hbar k. \tag{2.28}$$

After introducing the following operators to wave processes

$$\hat{E} = i\hbar\frac{\partial}{\partial t}, \quad \hat{p} = -i\hbar\nabla, \tag{2.29}$$

and including the potential energy $V(r)$ and new functions, depending on the wave $\vec{r}_v$ and diffusion $\vec{r}_d$ coordinates

$$\psi_{JS}(\vec{r} = \vec{r}_v + \vec{r}_d) = \psi_{JS}(\vec{r}_v, \vec{r}_d) = \psi_S(\vec{r}_v)\psi_J(\vec{r}_d), \tag{2.30}$$

we obtain the Schrödinger equation [1, 2] for bound states as follows:

$$i\hbar\frac{\partial\psi_S}{\partial t} = \frac{\hbar^2}{2m}\nabla^2\psi_S + V(r)\psi_S. \tag{2.31}$$

The free solutions in Eq. (2.5) do not satisfy the Schrödinger equation when $V(r) = 0$, and for a coincidence-free solution of (2.6), we must introduce the diffusion processes with the coordinates of different diffusing waves (2.30).

For bound systems, the coordinates of wave $\vec{r}_v$ and diffusion $\vec{r}_d$ coincide.

2.3 The Quantum Diffusion of an Electron in the Hydrogen Atom

The following quantum diffusion equation in a three-dimensional case can be obtained using (2.22):

$$\frac{\partial\psi_J}{\partial t} = \frac{\hbar}{4m}\nabla^2\psi_J. \tag{2.32}$$

Here, the diffusion coefficient depends only on mass and Planck's quantum constant $\hbar$, which is significant only for elementary particles such as electrons and protons [6]:

$$D_{C,e} = \frac{\hbar}{4m_e} = 0.2893\frac{cm^2}{s}, \quad D_{C,p} = \frac{\hbar}{4m_p} = 0.1575 \cdot 10^{-3}\frac{cm^2}{s}. \tag{2.33}$$

This diffusion happens for photons, free particles (2.6) and bound particles. In the center of forces and in region [2], where the kinetic energy $T = E - V(r)$ is negative, the wave function is decreasing, similar to $\exp[-r/R_n]$ with the decay length [2] obtained as follows:

$$R_n = k_n^{-1} = \frac{\lambda n}{2\pi} = \sqrt{\frac{2D_{Cn}}{\omega}} = \frac{\hbar n}{p} = \frac{\hbar n}{\sqrt{2m|E_n|}}. \tag{2.34}$$

For an electron bounded in the hydrogen atom [2, 3], we obtain a Bohr radius

$$R_n = \frac{\hbar^2 n^2}{k_e Z m e^2}, \quad n = 1, 2, 3, \ldots, \quad k_e = 9.10^9 \frac{N \cdot m^2}{C^2} \tag{2.35}$$

for a principal quantum number n and energy levels [2, A21], we have

$$E_n = -\frac{k_e^2 Z^2 m e^4}{2\hbar^2 n^2}. \tag{2.36}$$

The last formula can be obtained from (2.34) and (2.35). Substituting (2.36) in (2.34) and taking into consideration that $\hbar\omega = E_{n=\infty} - E_n$, we can find the quantum diffusion coefficient D_{Cn} and the connection between Bohr radius R_n and energy levels E_n for the hydrogen atom

$$D_{Cn} = \frac{\hbar}{4m} n^2 = D_C n^2, \quad R_n^2 |E_n| = 2\hbar D_{Cn} = 6.1 \cdot 10^{-39} n^2. \tag{2.37}$$

with the principal quantum number n. We have the smallest diffusion coefficient for the stable ground state when $n = 1$ and it is rapidly increasing for excited states like n^2. The obtained formula (2.37) can be used for the approximate evaluation of quantum systems: atoms, ions, molecules and point defect parameters in solids.

Now the radii of some atoms defined by the diffusion of electrons in physical vacuum according to the formula (2.37) can be calculated. Using the bound energies E_n of electrons [7] in the free atoms H, He, Li and Be in the external shells with energies 13.60, 39.47, 5.39 and 9.32 in eV, the radii of these atoms (in Angstroms, $10^{-10} m$) are obtained from (2.37) as follows:

$$R_{\text{H}} = 0.529\,\text{Å}, \quad R_{\text{He}} = 0.310\,\text{Å}, \quad R_{\text{Li}} = 1.68\,\text{Å}, \quad R_{\text{Be}} = 1.28\,\text{Å} \tag{2.38}$$

which were compared with the calculations provided in [8, 9]:

$$R_{\text{H}} = 0.529\,\text{Å}, \quad R_{\text{He}} = 0.31\,\text{Å}, \quad R_{\text{Li}} = 1.67\,\text{Å}, \quad R_{\text{Be}} = 1.28\,\text{Å}. \tag{2.39}$$

Taking into consideration the values obtained, we can suppose that the energies of the electrons in the external subshell define the atomic radii with a high accuracy and depend on the quantum diffusion coefficient D_C similar to some constant of physical vacuum for an electron and a square of a principal quantum number n.

Applying this conclusion, we will calculate the nuclear radii. For a deuteron, where a neutron and a proton are diffusing in the region $R_d = 2R$ of the physical vacuum defined by radius R in the coordinates of the center of mass, the formula (2.34) must be modified to obtain

$$R_d = 2R = \frac{\hbar}{\sqrt{2m|E_B|}}. \tag{2.40}$$

For deuteron, if the bound energy [10, 11] is $E_B = -2.225\,\text{MeV}$ and the reduced mass is

$$m = \frac{m_n m_p}{m_n + m_p}, \tag{2.41}$$

then we obtain $R = 2.158 \cdot 10^{-15}\,\text{m}$ or $2.158\,\text{fm}$. The charge radius [11] of a deuteron is 2.095 fm.

It is interesting to note that if the radius R is known, the energy of the bound system consisting of two equal particles from (2.37) for $n = 1$

$$E_B = \frac{-\hbar^2}{8mR^2} \tag{2.42}$$

can be determined. Using (2.42) we can find the Bohr's radii for the 1S states of charmed and bottom mesons using the charmed and the bottom quarks' masses and bound energies of quarks and antiquarks from the paper [6], A[20]

$$R_{C\bar{C}} = 0.144\,\text{fm}, \quad R_{b\bar{b}} = 0.08151\,\text{fm}, \tag{2.43}$$

$$R_{C\bar{C}} = 0.1704\,\text{fm}, \quad R_{b\bar{b}} = 0.05561\,\text{fm}. \tag{2.44}$$

For toponium [6] with ground state mass $M_t = 347.4\,\text{GeV}$ and top quark mass $m_t = 179.25\,\text{GeV}$, from (2.34) we obtain

$$R_{t\bar{t}} = 0.2212 \cdot 10^{-2}\,\text{fm}. \tag{2.45}$$

Taking the masses m_q of u and d quarks [6], $m_u = 0.35\,\text{GeV}$, $m_d = 0.35\,\text{GeV}$, and the proton root-mean-charge radius $R_p = 0.8621\,\text{fm}$ from (3.3), we obtain the bound energy of quarks and mass of a nucleon A[21] as follows:

$$E_B = 0.08416\,\text{GeV}, \quad M_n = 3m_q - E_B = 0.9658\,\text{GeV}. \tag{2.46}$$

Here, we consider that every quark is diffusing according to the mass center with reduced mass

$$m = \frac{0.5m_q m_q}{0.5m_q + m_q}. \tag{2.47}$$

The obtained mass of a nucleon is approximately equal to the mass of a neutron, i.e.,

$$m_n = 0.9396\,\text{GeV} \quad \text{and} \quad m_p = 0.9382\,\text{GeV}. \tag{2.48}$$

2.4   Solution of the Quantum Diffusion Equation for the Tunnel Effect for a Rectangular Barrier

From (2.32), separating the variable t in the expression of the wave function [2]

$$\psi_J(\vec{r}, t) = e^{\mp i\omega \cdot t}\psi_J(\vec{r}), \tag{2.49}$$

we obtain the quantum wave diffusion equations

$$\mp i\omega\psi_J(\vec{r}) = \frac{\hbar}{4m}\nabla^2\psi_J(r), \tag{2.50}$$

$$E\psi_J = \pm i\frac{\hbar}{4m}\nabla^2\psi_J, \quad E = \hbar\omega \tag{2.51}$$

for the free movement.

It is possible to present the quantum diffusion equation with potential $V(r)$

$$E\psi_J(\vec{r}) = \frac{i}{2}\left[-\frac{\hbar^2}{2m}\nabla^2\psi_J(\vec{r}) + V(r)\psi_J(\vec{r})\right]. \qquad (2.52)$$

This equation can be applied for the quantum diffusion in the case of bound states A[21, 22]. Now we will obtain the solution of the time-independent quantum diffusion equation for a rectangular well of infinite depth.

Introducing a rectangular barrier

$$V(0) = V(b) = V_0 \qquad (2.53)$$

and declaring the solutions of (2.52) for falling and reflected waves at the barrier $x_0 = 0$, we get

$$\psi_{J1} = A_1 e^{-kx+ikx} + B_1 e^{kx-ikx}. \qquad (2.54)$$

In the barrier region $0 \leq x \leq b$ of a small extension, we obtain the solution of (2.52) as follows:

$$\psi_{J1} = A_2 e^{-kx+ikx} + B_2 e^{kx-ikx}, \quad \kappa^2 = \frac{2m}{\hbar^2}V_0 - k^2, \quad k^2 = \frac{2mE}{\hbar^2}. \qquad (2.55)$$

The solution (2.54) for a transmitted wave at $x = b$ can be expressed in the following way:

$$\psi_{J3} = A_3 e^{(-k+ik)\cdot(x-b)}. \qquad (2.56)$$

From the equalities

$$\psi_{J1}(0) = \psi_{J2}(0), \qquad (2.57)$$

$$\frac{d\psi_{J1}}{dx} = \frac{d\psi_{J2}}{dx}, \quad x = 0, \qquad (2.58)$$

$$\psi_{J2}(x = b) = \psi_{J3}(x = b), \qquad (2.59)$$

$$\frac{d\psi_{J2}}{dx} = \frac{d\psi_{J3}}{dx}, \quad x = b, \qquad (2.60)$$

we obtain the following equations:

$$A_1 + B_1 = A_2 + B_2, \tag{2.61}$$

$$(-k + ik)A_1 + (k - ik)B_1 = (-\kappa + i\kappa)A_2 + (\kappa - i\kappa)B_2, \tag{2.62}$$

$$A_2 \exp[(-\kappa + i\kappa)b] + B_2 \exp[(\kappa - i\kappa)b] = A_3, \tag{2.63}$$

$$(-\kappa + i\kappa)A_2 \exp[(-\kappa + i\kappa)b + (\kappa - i\kappa)$$
$$B_2 \exp(\kappa - i\kappa)b = (-\kappa + i\kappa)A_3. \tag{2.64}$$

Solving the above equations, we obtain

$$\frac{A_3}{A_1} = \frac{4k\kappa \cdot \exp[(-\kappa + i\kappa)b]}{k^2 + \kappa^2 - (k - \kappa)^2 \exp[2(-\kappa + i\kappa)b]},$$
$$k^2 = \frac{2mE}{\hbar^2}, \quad \kappa^2 = \frac{2m}{\hbar^2}(V_0 - E) \tag{2.65}$$

or

$$\frac{A_3}{A_1} = \frac{4n \cdot \exp[(-\kappa + i\kappa)b]}{1 + n^2 - (n - 1)^2 \exp[2(-\kappa + i\kappa)b]}, \quad n = \frac{k}{\kappa},$$
$$n = \frac{k}{\kappa} = \sqrt{\frac{E}{V_0 - E}}. \tag{2.66}$$

Now, the transmission (diffusion) coefficient can be obtained as follows:

$$D = \frac{|A_3|}{|A_1|},$$
$$D = \frac{16k^2\kappa^2 \exp[-2\kappa b]}{s^2 - 2s\delta^2 \exp[-2\kappa b]\cos(\kappa b) + \delta^2 \exp(-4\kappa b)}, \tag{2.67}$$

where $s = k^2 + \kappa^2$, $\delta = k - \kappa$,
thus,

$$D = \frac{16n^2 \exp[-2\kappa b]}{N^2 + \Delta N^4 \exp[-4\kappa b] - N\Delta N^2 \cos[2\kappa b] \exp[-2\kappa b]}, \tag{2.68}$$

where $N = 1 + n^2$, $\Delta N = n - 1$.

Taking an approximate value of the denominator, where the width b of the barrier is large compared to the wave length $\lambda = 2\pi/\kappa$,

$\kappa b \gg 1$, we get

$$D \cong D_0 \exp[-2\kappa b], \quad D_0 = \frac{16n^2}{(1+n^2)^2}, \tag{2.69}$$

which is the same expression as in Ref. [2]. For this case, from (2.68) we can get the equivalent expression [10] as follows:

$$D = \frac{16k^2\kappa^2 \exp[-2\kappa b]}{(k^2 + \kappa^2)^2}. \tag{2.70}$$

For the obtained solution, we have essential differences of the wave functions in the barrier region where we have the oscillating function (2.55) and the essentially different classical solution [2]

$$\psi_{J2} = A_2 e^{-\kappa \cdot x} + B_2 e^{\kappa \cdot x}, \quad \kappa^2 = \frac{2m}{\hbar^2}V_0 - k^2, \quad k^2 = \frac{2mE}{\hbar^2}. \tag{2.71}$$

2.5   Conclusions

The Heisenberg uncertainty principle related to the diffusion equation is a powerful method for explaining the differences between classical and quantum physics and expanding the theoretical and practical applications. From (2.26), we infer that the frequency of disappearance of an electron in the hydrogen atom is $1/\Delta t = 0.156 \cdot 10^{22}\,\text{s}^{-1}$, which occur by annihilation and interactions with virtual positrons in physical vacuum. They can appear at the distance $c\Delta t = 0.192 \cdot 10^{-2}\,\text{Å}$. In this case, the electrons cannot move in circular Bohr orbits. Free movement of quantum particles is a stochastic process and is correlated with quantum diffusion in physical vacuum (2.33) with decreasing stochastic wave packets (2.9). The formulas obtained ((2.34) and (2.35)) can be used for evaluating the radii of spherical defects in solids. This is important for the analysis of point defects by X-ray diffraction experiments A[23] and for application to new superdiffusion technologies for the production of semiconductors A[27]. The wave functions in the potential region (2.55) are oscillating in a way that is similar to the free movement represented by Eqs. (2.16) and (2.17). In the potential barrier region, we cannot use the assumption that we have not considered the scattering of a real particle, but there are virtual processes [14] quantum diffusion of a particle that are involved in the physical vacuum. This comment is interesting because in paper [15] the hypothesis that a particle with mass m takes part in Brownian motion with diffusion coefficient $\hbar/2m$ obtained from the Newton's second law of motion was applied. The wave functions of particles having continuous trajectories [14] cannot be used to describe the quantum states.

References

1. Sokolov, A. A. and Tumanov V. S. (1956). *J. Exp. Theoret. Phys.*, **30**, p. 802 (in Russian).
2. Bochm, D. (1989). *Quantum Theory*, Dover Publications, Inc., New York, p. 672.
3. Sokolov, A. A. and Ternov I. M. (1970). *Quantum Mechanics and Atomic Physic*, Prosveshchenie, Moscow, p. 423 (in Russian).
4. Glicksman, M. E. (2000). *Diffusion in Solids: Field Theory, Solid-State Principles, and Applications*, A Wiley-Interscience Publication, John Willey & Sons, Inc., New York, p. 498.
5. Kotelnikov, V. A. (2009). *Adv. Phys. Sci.*, **179**(2), p. 204 (in Russian).
6. Griffiths, D. (2008). *Introduction to Elementary Particles*, Wiley-Wch Verlag GmbH&Co, KGaA, Weinheim, p. 470.
7. Karazija, R. (1987). *The Theory of X-ray and Electronic Spectra of Free Atoms. An Introduction*, Mokslas, Vilnius, p. 314 (in Russian).
8. Drake, G. W. (1999). High precision calculation of atomic helium, *Phys. Sci.*, **83**, 82–92.
9. Roetti, C. and Clementi, E. (1974). Simple basis sets for molecular wave functions containing atoms from $Z = 2$ to $Z = 54$, *J. Chem. Phys.*, **60**, 4725.
10. Blatt, J. M. and Weisskopf, V. F. (1991). *Theoretical Nuclear Physics*, Dover, New York, p. 864.
11. Ericson, T. E. O. and Mahalanabis, J. (1985). *Nuclei*, **322**(2), 237–239.
12. Rosenfelder, R. (2000). *Physics Letters B.* **479**, 381.
13. Nordlund, K., Partyka, P. and Averback, R. S. (1997). Fully atomistic analysis of diffuse X-ray scattering spectra of silicon defects, *Mat. Res. Soc. Symp. Proc.*, **469**, 199–204.
14. Veltman, M. (2003). *Facts and Mysteries in Elementary Particle Physics*, World Scientific, Singapore, p. 348.
15. Nelson, E. (1966). Derivation of the Schrodinger equation from Newtonian mechanics, *Phys. Rev.*, **150**, 1079.

Chapter 3

Nuclear Forces

3.1 The Interactions between Nucleons

Gluons and quarks are confined in a limited space, and the forces in nuclei are produced due to the interchange of mesons whose field is similar to the electromagnetic field, yet is of a much shorter range. Keeping in mind the above fact, we will calculate the interaction radius of nucleons in nuclei. For the $^{197}_{79}Au$ nucleus, both neutrons and protons are diffusing in the region $R_d = 2R$ of physical vacuum defined by radius R in the coordinates of the center of mass. The formula A[21] for calculating the distance of interaction of an external proton with bound energy 7.73 MeV can be given as follows:

$$R_d = 2R = \frac{\hbar}{\sqrt{2m|E_B|}} = 1.638 \cdot 10^{-15}\,\text{m} = 1.638\,\text{fm}. \qquad (3.1)$$

Yukawa declared that the interaction with a short interaction radius was generated for field quanta with negligible rest mass. The electromagnetic field is occupied by photons of zero rest mass that satisfy the wave equation

$$\left(\nabla^2 - \frac{1}{c^2}\frac{\partial^2}{\partial t^2} \right) U(r, t) = 0, \qquad (3.2)$$

which can be written as

$$-p^2 + \frac{1}{c^2}E^2 = 0. \qquad (3.3)$$

The corresponding relativistic equation for particles having the rest mass is

$$-p^2 - m^2c^2 + \frac{1}{c^2}E^2 = 0. \tag{3.4}$$

After the introduction of the quantum mechanics operators

$$\vec{p} = -i\hbar\vec{\nabla}, \quad E = i\hbar\frac{\partial}{\partial t}, \tag{3.5}$$

we can transform it into the below Klein–Gordon equation for a spinless relativistic free particle

$$\left(\nabla^2 - \frac{m^2c^2}{\hbar^2} - \frac{1}{c^2}\frac{\partial^2}{\partial t^2}\right)\phi = 0. \tag{3.6}$$

This wave equation for the time-independent meson field

$$(\nabla^2 - \mu^2)\phi = 0, \quad \mu = \frac{mc}{\hbar} \tag{3.7}$$

was solved similar to the following the spherically symmetric solution

$$\phi = g\frac{e^{-\mu \cdot r}}{r}, \tag{3.8}$$

where g is an undetermined constant which plays the same role as the charge in the electrostatic case. In this case, it depends on the source at the origin but in a different way, where the mass and binding energy of the nucleon must be included (3.1). The potential energy between two electrical charges is $V_C = qU = q^2/4\pi \cdot r$. Then, [1] can be written similarly for the nucleon interaction potential of strength g

$$V(r) = g\frac{e^{-\mu \cdot r}}{r}, \tag{3.9}$$

arising from the continuous transfer of virtual mesons of rest mass m between two nucleons.

The interactions can be classified according to the value of a characteristic dimensionless constant related through a coupling constant. The strong interactions increase the interaction cross section and reduce the interaction time [2, 3]. The interaction strength gives a dimensionless constant for the strong interactions, i.e., $g^2/\hbar c \approx 1, \ldots, 10$. For the electromagnetic interaction, $e^2/\hbar c = 1/137$. The weak interaction has the potential shown in (3.9) and

$g_w^2 \approx 10^{-5}$. The gravitation interaction between two protons can be represented by the constant [3] $Gm_p/\hbar c \approx 6.0 \cdot 10^{-39}$, which is so small that it can be usually neglected in particle physics. The duration of expedition of virtual mesons from nucleon to nucleon can be found from the uncertainty principle

$$\Delta t \Delta E \approx \hbar, \quad \Delta E \approx mc^2, \quad \Delta r \approx c\Delta t \approx \frac{\hbar}{mc}, \tag{3.10}$$

and the range of force Δr is equal to the Compton wave length. Calculating the range of potentials given in (3.9) arising from the different types of meson exchange and considering

$$\mu \cdot r_0 = 1, \quad r_0 = \frac{1}{\mu} = \frac{\hbar}{mc}, \tag{3.11}$$

we obtain a result similar to Δr in (3.10). Considering the mass [2] as $135\,\text{MeV}/c^2$ for the lightest pi meson π^0, from (3.11) we obtain the range for the force between nucleons as 1.429 fm. The calculations (3.1) using mass and binding energies of nucleons, diffusing in excited physical vacuum, give a similar result, i.e., 1.638 fm. However, the result in formula (3.1) depends on the quantum states of sources of virtual mesons that are in the nuclei. More massive virtual mesons [3] $(\pi, \sigma, \rho, \omega)$ with spins and isospins (J^π, T) are described by the different ranges and kinds of potential, which can also play an important part in the interactions that occur between nucleons. The term virtual indicates that the meson cannot be released from the nucleus when the energy is less than $m_0 c^2$. Emission of virtual mesons and re-capture must occur under the violation of conservation of energy within a short time interval Δt. The resulting potential between two nucleons [2] depends on the spin, i.e.,

$$V(\vec{r}) = V_C(r) + V_{C \cdot S}(r)\frac{\vec{S}_1 \cdot \vec{S}_2}{\hbar^2} + V_T(r)\frac{\hat{S}_{12}}{\hbar^2} + V_{SL}(r)(\vec{L} \cdot \vec{S}). \tag{3.12}$$

The potential presented is invariant with regard to the rotations, mirror reflections, and interchange of the spins of the interacting nucleons. For this case, the states of interacting nucleons can be classified using quantum numbers $\vec{J} = \vec{L} + \vec{S}$ representing the squares of the sum of orbital momentum $\vec{L}$ and total spin $\vec{S} = \vec{s}_1 + \vec{s}_2$ moments, their projection M_J, the sum of spins S and the parity $(-1)^L$. The total orbital quantum number L can be used when the tensor of non-central forces is not included for singlet states of $^{2S+1}L_j$, $S = 0$.

Here, we have $V_C(r)$, the spin-independent central potential; $V_{CS}(r)$, the spin-dependent central potential, $V_T(r)$, tensor potential, V_{LS}, the two-body spin–orbit potential, and $\vec{s} = \frac{1}{2}\hat{\sigma}\cdot\hbar$, where $\vec{\sigma}$ is the Pauli spin matrix. The non-central interaction represents the tensor potential which depends on the angles between the spin vectors of the neutron and the proton and the radius vector. Tensor forces are proportional to the below non-central tensor operator

$$\hat{S}_{12} = \frac{3}{r^2}(\vec{\sigma}_1 \cdot \vec{r})(\vec{\sigma}_2 \cdot \vec{r}) - \vec{\sigma}_1 \cdot \vec{\sigma}_2, \tag{3.13}$$

which is equal to zero for the singlet state $S = 0$. The tensor potential $\hat{S}_{12}$ can explain the electric quadruple moment of the deuteron described by mixed triplet states, $a \cdot \psi(^3S_1) + b \cdot \psi(^3D_1)$. For the normalized function of triple states [1] $a^2 + b^2 = 1$, $b^2 = 0.004$, influence of state $^3D_1$ is small, about 4%. The spin–orbit potential V_{SL} can be expressed as follows:

$$V_{SL} = V_{SL}(r)(\vec{L} \cdot \vec{S}), \, (\vec{L} \cdot \vec{S})$$
$$= \frac{1}{2}[J(J + 1) - L(L + 1) - S(S + 1)]. \tag{3.14}$$

This last expression is equal to zero when $S = 0$, $(L = J)$ or $L = 0$, $(S = J)$. Using the total spin dependence and tensor terms, (3.12) can be rewritten [2] as follows:

$$(\vec{s}_1 \cdot \vec{s}_2) = \frac{1}{4}[2S(S + 1) - 3], \tag{3.15}$$

$$\hat{S}_{12} = 6(\vec{n}\vec{S})^2 - 2\vec{S}^2. \tag{3.16}$$

From the above formulas we see that the spin-dependent expectation values of $(\vec{s}_1 \cdot \vec{s}_2)$ and the tensor force operators are different for singlet $S = 0$ and triplet $S = 1$ states. The operator $\hat{S}_{12}$ depends on the angle between the unit vector $\vec{n}$ directed from one nucleon to another and $\vec{S}$. The wave function Y_{lSJM} that depends on the eigenfunctions of the orbital moment $Y_{lm}(\vartheta, \varphi)$ and the spin moment $\chi_{S\mu}$ as follows:

$$Y_{lSJM} = \sum_{m+\mu=M} (lmS\mu|JM)Y_{lm}\chi_{S\mu} \qquad (3.17)$$

is expressed by Clebsch–Gordan coefficients. Interaction of the potential (3.12) and energy of the two potentials depend on the quantum states ISJ. The wave function of the deuteron representing the bound states of both neutron and proton is a linear combination of the following functions [2] that are expressed by orthonormal functions Y_{lSJM}.

$$\psi(^3S_1) = \frac{u(r)}{r}Y_{011M}, \quad \psi(^3D_1) = \frac{\omega(r)}{r}Y_{211M} \qquad (3.18)$$

Then the required normalization of the deuteron function

$$\int |\psi_0|^2 d\vec{r} = 1 \qquad (3.19)$$

can be obtained by using

$$\int_0^\infty (u^2 + \omega^2)dr = 1 \qquad (3.20)$$

and the probabilities of deuterons in states $^3S_1$ and $^3D_1$, i.e.,

$$P(^3S_1) = \int_0^\infty u^2 dr, \;\; P(^3D_1) = \int_0^\infty \omega^2 dr, \;\; P(^3S_1) + P(^3D_1) = 1. \qquad (3.21)$$

The Schrödinger equation for the neutron–proton bounded system or the deuteron [2]

$$[-(\hbar^2/M)\Delta + V_C + V_T S_{12} - E]\psi = 0 \qquad (3.22)$$

was solved for the function $\psi_d = a \cdot \psi(^3S_1) + b \cdot \psi(^3D_1)$ only at large distances for the spin function $r > r_0$, when the total spin coincides with the projection of the total moment of the deuteron. The asymptotes of solutions are expressed by the binding energy E, i.e.,

$$u(r) = C_U \exp[-\alpha \cdot r], \tag{3.23}$$

$$w(r) = C_W \left[1 + \frac{3}{\alpha r} + \frac{3}{(\alpha r)^2}\right] \exp[-\alpha \cdot r], \tag{3.24}$$

$$\alpha = \sqrt{2ME}/\hbar, \quad E = 2.226 \, \text{MeV}.$$

The range of central (3.22) and tensor forces can be approximately expressed by the radius of the quantum diffusion A[21] region of the nucleus, $R = \hbar/\sqrt{2ME}$ which is defined as $R = 2.158 \cdot 10^{-15}$ fm for deuterons or as 2.158 fm for the interaction energy E of nucleons.

At small distances, the asymptotic solutions are $u \cong r$ and $w \cong r^3$ and also r^2 in Ref. A[19] for $l = 0$, when the relativistic correction for kinetic energy is included. The component of the quadruple moment operator for the deuteron in the case $M = J = 1$ is given as

$$Q = \frac{1}{4}e(3z^2 - r^2) \tag{3.25}$$

and can be used for the calculation of the expectation value, i.e.,

$$\langle Q \rangle = \int_0^\infty \psi_d^* Q \psi_d \, dr, \tag{3.26}$$

which for $P(^3D_1) = 0.04$ gives [1, 2] $\langle Q \rangle = 2.82 \cdot 10^{-27} \text{e} \cdot \text{cm}^2$.

3.2   The Shell Model and Mean Field Potentials

The simplest shell model of hydrogen atom was proposed by Niels Bohr by introducing the quantization of angular momentum

$$mv \, r = l\hbar, \quad l = 1, 2, 3, \ldots \tag{3.27}$$

for an electron moving with velocity v at a distance r from the positive proton or any nucleus with charge eZ. For a circular orbit the centrifugal forces are equal to the attractive forces as shown below:

$$\frac{mv^2}{r} = \frac{Ze^2}{r^2}. \tag{3.28}$$

Then energy of an electron can be calculated using the above equations as follows:

$$E_n = \frac{mv^2}{r} - \frac{Ze^2}{r^2} = -\frac{m}{2} \cdot \frac{Z^2 e^4}{\hbar^2 n^2}, \quad n = l = 1, 2, 3, \ldots . \quad (3.29)$$

Using quantum diffusion of bounded particles from the decreasing wave functions, we obtain the radius A[21] of the moving electrons [3] in the hydrogen atom with energies E_n as follows:

$$R_n = \hbar n / \sqrt{2m|E_n|} = \frac{\hbar^2 n^2}{mZe^2}. \quad (3.30)$$

The more exact result can be obtained by solving the time-independent Schrödinger equation [3] for hydrogen atom $Z = 1$ or ion $Z > 1$ with one electron as follows:

$$-\frac{\hbar^2}{2\mu} \nabla^2 \psi(\vec{r}) + V(r)\psi(\vec{r}) = E\psi(\vec{r}), \quad V(r) = -\frac{Ze^2}{r} \quad (3.31)$$

after separation of the radial angle r and the variables ϑ, φ, we have

$$\psi(\vec{r}) = \frac{u_{nl}(r)}{r} Y_{lm}(\vartheta, \varphi), \quad l = 0, 1, \ldots, n-1, \quad m = l, l-1, \ldots, -l, \quad (3.32)$$

requiring that the spherical harmonics Y_{lm} and radial wave functions u_{nl} provided in the following equation

$$-\frac{\hbar^2}{2\mu} \frac{d^2 u_{nl}}{dr^2} + \frac{\hbar^2 l(l+1)}{2\mu} + V(r)u_{nl} = Eu_{nl}, \quad \mu = \frac{mM}{m+M} \approx m, \quad (3.33)$$

must be normalized to unity

$$\int Y_{lm}^* Y_{lm} d\Omega = \int_0^\infty u_{nl}^* u_{nl} dr = 1. \quad (3.34)$$

Using the asymptotic behavior of solutions of Eq. (3.33) at infinity and at zero for r, we have

$$u_{nl,\infty} = e^{-kr}, \quad u_{nl,0} \to r^{l+1}, \quad k = \sqrt{\frac{-2\mu E}{\hbar^2}}, \quad E < 0, \quad (3.35)$$

and representing $u_{nl} = e^{-kr}u_{nl,0}$, we can rewrite the above equation as follows:

$$\frac{d^2 u_{nl,0}}{dr^2} - 2k\frac{du_{nl,0}}{dr} - \frac{l(l+1)}{r^2}u_{nl,0} + \frac{2\mu Ze^2}{\hbar^2}u_{nl} = 0. \tag{3.36}$$

Substituting the solution of the last equation similar to the power series

$$u_{nl,0} = \sum_{N=0}^{\infty} a_N r^{l+1+N}, \tag{3.37}$$

we obtain the recursion relation

$$a_{N+1} = \frac{-2\mu Ze^2/\hbar^2 + 2k(l+1+N)}{(l+N+2)(l+N+1) - l(l+1)}a_N. \tag{3.38}$$

The series (3.38) must be terminated and it is essential that their degree must be N with a_{N+1}. In this case, we obtain

$$k = \frac{\mu Ze^2}{\hbar^2 n}, \quad E_n = -\frac{\mu Z^2 e^4}{2\hbar^2 n^2}, \quad n = N + l + 1,$$

$$n = 1, 2, 3, \ldots, \quad l = 0, 1, 2, \ldots, n-1, \tag{3.39}$$

where we define the principal quantum number in n and orbital momentum $l = 0, 1, 2, 3, \ldots$ which can be marked as $s, p, d, f, g, h, \ldots$. The quantum states are additionally defined by orbital momentum projection or magnetic quantum numbers m, $m = l, l - 1, \ldots, -l$. Here, the increasing energies have infinitely many discrete energy levels till $E_n = 0$. The ground state corresponds to $n = 1, l = 0$ with the energy level $E_1 = -13.6 \, \text{eV}$. By including the two spin projections defined by quantum numbers $\pm\frac{1}{2}$ and by using Eq. (2.13), we obtain the number of independent stationary states of hydrogen as follows:

$$2\sum_{l=0}^{n-1}(2l+1) = 2n^2, \quad n = 1, 2, 3, \ldots. \tag{3.40}$$

The electrons in the states and shells with definite values n have $2n^2$ degenerate states with the same definite energies. According to the Pauli's principle, only one fermion particle can be present in a specific quantum state defined by a specific quantum number.

The states nl filed by the $2(2l+1)$ electrons represent closed subshells. The degenerated energy levels with the same n are equal and this exists only for Coulomb potentials (3.31).

The nuclear mean field is essentially different from the Coulomb potential and has strong spin–orbit coupling (3.14). The quantum diffusion region of neutrons and protons is localized in a significantly smaller sphere [2] and the interaction between nucleons occurs only at small distances (3.1) $r = 1.5\,\text{fm}$. Therefore, the shell structure in the nucleus is very different from the shell structure in the atom. The nuclear stability, binding energies, and spins depend on the number of protons Z and neutrons N. In the same shells, energy levels are closer than the energy intervals between the neighboring shells. The nucleons in closed shells with the magic number [2] of protons $Z = 2, 8, 20, 28, 40, 50, 82, (126)$ or 164 and neutrons $N = 2, 8, 20, 28, 50, 82, 126, 184, 196, (272), 318$ couple up to spin equal to zero and have increased the stability and the total abundance of isotopes on the Earth. The nucleons have very little interaction at the ground level of nuclei. The independent particle model in one nucleon approximation is the most prominent in the shell model calculations of energies at the ground and higher [2] levels of excited states of nuclei. The short-range interaction between nucleons can be approximately replaced by the movement of the nucleons in the average Woods–Saxon potential [2]:

$$V(r) = -V_0 \frac{1}{1 + e^{(r-R)/a}} + U_{sl}(r), \quad V_0 \approx 60\,\text{MeV},$$

$$R = 1.25 A^{1/3}\,\text{fm}, \quad a = 0.65\,\text{fm},$$

$$U_{sl}(r) = -\kappa \frac{\partial V(r)}{\partial r} \cdot \frac{1}{r}(\vec{s}\vec{l}), \tag{3.41}$$

$$(\vec{s}\vec{l}) = \frac{1}{2}l, \quad j = l + 1/2, \tag{3.42}$$

$$(\vec{s}\vec{l}) = -\frac{1}{2}(l+1), \quad j = l - 1/2 \tag{3.43}$$

with the added spin–orbit interaction $U_{sl}(r)$. The last two equations are obtained from (3.14). The wave function can now be expressed as follows:

$$\psi_{nljm}(r) = \frac{u_{nlj}(r)}{r} Y_{ljm}(\vartheta, \varphi), \qquad (3.44)$$

$$Y_{ljm} = \sum_{m_l, m_s} \langle lm_l s m_s | jm \rangle Y_{lm}(\vartheta, \varphi) \chi_{1/2}(m_s). \qquad (3.45)$$

The Saxon–Woods potential (3.41) represents the average nucleons' interaction and can be defined by scattering experiments [5], their A[1, 2], and calculations of nuclear energy levels A[14]. The fluctuations about this average potential are small because the average separation between nucleons in a nucleus is sufficiently large ≈ 2.4 fm. However, many investigations show that calculations of excited states and reactions are not exact without including the residual interactions [6]. The energies of the excited states can be significantly improved by including a relativistic corrections for masses A[13]. The Woods–Saxon potentials can be investigated by the optical model [6] applications for elastic and non-elastic scattering in the Woods–Saxon potentials, representing one-particle shell models. The results of the quantum mechanical studies can be improved by investigating the elastic scattering and absorption by including the additional imaginary potential $iW(r)$. A complex potential means that the wave function and wave vector of the particles scattering in the nuclei are complex

$$\psi = \exp(iKz), \quad K = K_r + iK_m, \qquad (3.46)$$

and describe some part of its absorption

$$|\psi^* \cdot \psi| = \exp(-2K_m z). \qquad (3.47)$$

The real part of the optical potential for nucleons scattering on nuclei can be defined by the density of the target nucleons, $\rho(r_t)$:

$$V(\vec{r}) = \int \rho(\vec{r}_t) V(\vec{r} - \vec{r}_t) d^3 \vec{r}_t. \qquad (3.48)$$

Considering the experimental results [5], it was seen that the real part $V(r)$ of optical potential is about 10 times larger than the imaginary part $W(r)$. The weak absorption and large mean free part (about

10 fm) [6] were obtained for low-energy neutrons $E \ll V$ scattering

$$\lambda = \frac{1}{2K_m}, \quad K_m = \frac{W}{\hbar}\sqrt{\frac{m}{2V}}. \tag{3.49}$$

The results show that the model for the independent particles in the nuclei is sufficiently realistic and the Woods–Saxon potential with spin–orbit interactions defined by optical model calculations can be used. These calculations can be significantly improved using relativistic corrections for nucleon masses in optical model calculations A[3] and semi-relativistic model of nuclei A[3, 14, 15]. The more exact semi-relativistic model allows to define experimental one-particle energy levels for many nuclei A[15] changing only the depths for the Woods–Saxon potential proposed by Chepurnov. This fact underlines the importance of the semi-relativistic model where relativistic corrections A[3] for nucleons' masses are included. These negative corrections for levels near the ground state are significant A[15] and important in the optical model calculations and increase the stability of large nuclei A[16, 17]. The exact calculations of these corrections require the application of integral equations with potential representation A[8] using the harmonic oscillator potential for modelling the following equation

$$V(r) = -V_0 + \frac{1}{2}m\omega^2 r^2 \tag{3.50}$$

and also for modelling the physical and nonphysical solutions with it. Here, V_0 is the depth of the potential well, m is the mass of the nucleon and ω is the frequency of oscillations. This simple model presents realistic results, when spin–orbit interaction is included A[8] for energy levels of neutrons and magic numbers 2, 8, 20, 40 of light nuclei, when $\omega = \omega_0$, i.e.,

$$\omega_0 = 41A^{-1/3} \,\text{MeV}. \tag{3.51}$$

For this reason, the harmonic oscillator potential can be used as a model potential for multiplicative perturbation theory A[8] applied for relativistic corrections to calculations of one-nucleon levels A[14]. Here, the Woods–Saxon potential used is the average between finite and harmonic oscillator potentials. The decreasing region of the

Woods–Saxon potential depends on the nuclear reactions and excited states which can be calculated by only including the relativistic corrections A[13, 14] for the nucleon masses. The depths of the Woods–Saxon potentials for protons are smaller in the presence of the Coulomb interaction, i.e.,

$$V_z(r) = \frac{(Z-1)e^2}{r}\left(\frac{3}{2} - \frac{1}{2}\left(\frac{r}{R}\right)^2\right), \quad r < R,$$

$$V_Z(r) = \frac{(Z-1)e^2}{r}, \quad r > R. \tag{3.52}$$

As a result of spin–orbit interactions (3.41), the energy levels for the levels with the same quantum number nl for (3.42) $j = l + 1/2$ reduce with the increasing binding energies and for (3.43) $j = l - 1/2$ increase with the decreasing binding energies. The lowest states $1n_{1/2}$ in the nuclei have two neutrons with spin projections $m_S = \pm 1/2$ and in the same upper states have two protons. In the nucleus of helium, we have magic numbers of neutrons $N = 2$ and protons $Z = 2$. The second upper shell $1p$ with 6 nucleons in the nuclei consists of state $1p_{3/2}$ with $2j + 1 = 4$ nucleons and state $1p_{1/2}$ with $2j + 1 = 2$ nucleons. In this case, we can obtain the ^{16}O nucleus with the magic number of 8 protons and neutrons. The following shells will be $1d$ with 10 nucleons and $2s$ with 2 nucleons which consist of the states $1d_{5/2}$, $1d_{3/2}$ and $2s_{1/2}$. Here we have the following magic numbers for neutrons $N = 20$ and protons $N = 20$. Similarly [6], A[14] upper nuclei shells $N, Z28, 50, 82, N126$, and A[15] $Z114, 126, N126, 184, 214$ can be considered.

In the single-particle shell model, the residual remaining interaction is the pairing interaction. Then the spins and parities J^π can be defined by keeping in mind that nucleon spins in closed shells couple up to spin equal to zero. Also, for ground states for even–even neutrons and protons, we have spin equal to zero and positive parity. For a nucleus $^{17}_{8}O$ with one neutron outside the closed shell $1p$, the spin and parity are defined by the neutron state $1d_{5/2}$. For the nucleus with unpaired neutron $^{15}_{8}O$, we have a neutron hole in state $1p_{1/2}$. In odd–odd nucleus, we have j_n spin for neutrons and j_p

spins for protons, thus the spin of the nucleus can be between $|j_n - j_p|$ and $(j_n + j_p)$.

The wave function in the single-particle shell model can be presented like a product of one-particle wave functions

$$\psi_t = \text{antisymmetric } \psi_1 \psi_2 \psi_3 \cdots \psi_n. \qquad (3.53)$$

The parities of the nucleus in this case are equal to the parities of single–wave functions. Therefore, we have positive parities for even–even nuclei. For odd nucleus, we have the parity similar to the odd nucleon's wave function. For example, parity of the $^{17}_{8}O$ nucleus is positive and is defined by the odd neutron state $1d_{5/2}$. For even–even nuclei, the magnetic dipole moment $\vec{\mu}$ is zero as the nucleus spin is zero. For odd number of nucleons, $\vec{\mu}$ is defined by the unpaired nucleon. If it is a neutron, then it is only a spin contribution to the magnetic moment. The proton gives orbital and spin contribution proportional to the nuclear magneton μ_N, i.e.,

$$\vec{\mu} = \vec{\mu}_l + \vec{\mu}_s = g_l \mu_N \vec{l} + g_s \mu_N \vec{s}, \quad \mu_N = e\hbar/2m, \qquad (3.54)$$

where $g_l = 1$ and $g_s = 5.586$ for a proton and $g_l = 0$ and $g_s = -3.826$ for a neutron. $\vec{l}$ and $\vec{s}$ are expressed in units of $\hbar$.

In the heavier nuclei, the shell model in the one-particle approach cannot predict the magnetic moments and energy levels of nuclei correctly. The semi-relativistic one-particle model with the relativistic corrections included A[3] gives the correct one-particle displacing states A[15] and almost accurate calculated energy levels A[14] for many nuclei. The more exact Hamiltonian of a nucleus must include one-particle and two-particle elements

$$\widehat{H} = \sum_i \widehat{H}_0(\vec{r}, \vec{p}_i) + \sum_{i,j} V(|\vec{r}_i - \vec{r}_j|). \qquad (3.55)$$

Here, the first term represents the one-particle shell model which we improved by including the relativistic corrections A[15]. The second term is represented by the complicated Hamada–Johnston potential [7] with infinite repulsion at the radius 0.49 fm. The interacting nucleons cannot be closer than this distance and must satisfy the Pauli's exclusion principle. These reasons generate pairing forces for nucleon moments and saturation of the nuclear forces and allow to use the independent-particle shell model as a good approximation. The Woods–Saxon potentials with relativistic corrections represent the main term of the Hamiltonian $\hat{H}_0$, and the second term is smaller and cannot be included in the calculations for nuclei with shells near magic numbers.

When the external shells have about half-filled subshells and the nuclei are nonspherical, the second term of (3.55) must be included for evaluating the residual interactions. For the evaluation of matrix elements, the two-particle potential must be expanded in following way:

$$V(|\vec{r}_i - \vec{r}_j|) = \sum_l V_l(r_i, r_j) P_l(\vartheta_{ij}) \tag{3.56}$$

or

$$V(|\vec{r}_i - \vec{r}_j|) = \sum_l V_l(r_i, r_j) \sum_m Y_{lm}^*(\vartheta_i \phi_i) Y_{lm}(\vartheta_j \varphi_j). \tag{3.57}$$

If the interactions take place only on the surface of the nucleus, we can separate the radial variables using the approach [6] for practical calculations, i.e.,

$$V_l = \text{const} \cdot r_i^l \cdot r_j^l \tag{3.58}$$

or use the multiplicative perturbation theory A[8] for non-spherical potential $V(r, \vartheta, \varphi)$.

References

1. Enge, H. A. (1966). *Introduction to Nuclear Physics*. Addison-Wesley Publishing Company, Ontario, p. 582.
2. Devanarayanan, S. (2016). *A Text Book on Nuclear Physics*, Lexington, KY.
3. Yung-Kuo Lim. (ed.) *Problems and Solutions on Atomic, Nuclear and Particle Physics*, World Scientific, New Jersey, 2003, p. 717.
4. Bohm, D. (1989). *Quantum Theory*, Dover Publications, Inc., New York, p. 646.
5. Strzalkowski, A. (1978). *Introduction to Physics of Atomic Nucleus*, Scientific State Publishing, Warsaw, p. 637 (in Polish).
6. Jelley, N. A. (1990). *Fundamentals of Nuclear Physics*, Cambridge University Press, Cambridge, p. 278.
7. Heyde, K. L. G. (1994). *The Nuclear Shell Model*, Springer-Verlag, Berlin, p. 438.

Chapter 4

Systems of Micro Particles

The wave function's dependence on time and the coordinates of particles can be found when solving the following common Schrödinger equation:

$$i\hbar \frac{\partial \psi(x_1, x_2, \ldots, x_N, t)}{\partial t} = \hat{H}\psi(x_1, x_2, \ldots, x_N), \qquad (4.1)$$

with the Hamiltonian

$$\widehat{H} = \sum_{j}^{N} \hat{T}_j + \sum_{j}^{N} V_j + \sum_{j \neq k}^{N} V_{jk}, \quad \hat{T}_j = \frac{\hbar^2}{2m_j}\nabla_j^2, \qquad (4.2)$$

which consists of kinetic operators $\hat{T}_j$, potential energy $\hat{V}_j$ and interaction between particles V_{jk}. Substituting

$$\psi = e^{-\frac{i}{\hbar}Et}\varphi \qquad (4.3)$$

into (4.1), we obtain the time-independent Schrödinger equation

$$\hat{H}\varphi(x_1, x_2, \ldots, x_N) = E\varphi(x_1, x_2, \ldots, x_N), \qquad (4.4)$$

whose solutions for the obtained time-independent Schrödinger equation connect the set of eigenfunctions φ_α and eigenvalues E_α in bounded stationary systems. The quantum numbers and states of wave functions define the eigenvalues of the commuting operators with the Hamiltonian. Interactions between negatively charged electron $-e$ and positively charged Ze nucleus of ions with atomic number Z in the hydrogen atom in the spherically symmetric Coulomb field are included in the Hamiltonian

$$-\frac{\hbar^2}{2m}\left[\frac{1}{r^2}\frac{\partial}{\partial r}\left(r^2\frac{\partial}{\partial r}\right)+\frac{1}{r^2}\hat{\Delta}\right]\psi(r,\vartheta,\varphi)$$

$$-\frac{Ze^2}{r}\psi(r,\vartheta,\varphi)=E\psi(r,\vartheta,\varphi), \tag{4.5}$$

where the Hamiltonian commute $\hat{H}\cdot\hat{l}^2\psi=\hat{l}^2\hat{H}\psi$ with the orbital moment square $\hat{l}^2$ and projection $\hat{l}_z$ operators

$$\hat{l}^2=-\hbar^2\left(\frac{1}{\sin\vartheta}\frac{\partial}{\partial\vartheta}\left(\sin\vartheta\frac{\partial}{\partial\vartheta}\right)+\frac{1}{\sin^2\vartheta}\frac{\partial^2}{\partial\varphi^2}\right)=-\hbar^2\hat{\Delta}, \tag{4.6}$$

$$\hat{l}_z=-i\hbar\frac{\partial}{\partial\varphi}. \tag{4.7}$$

Solution of (4.5) can be expressed by separating the radial variables r and the angles ϑ,φ

$$\psi(r,\vartheta,\varphi)=\frac{u(r)}{r}Y_{lm}(\vartheta,\varphi), \tag{4.8}$$

where spherical harmonics $Y_{lm}(\vartheta,\varphi)$ are eigenfunctions of operators $\hat{l}^2$ and $\hat{l}_z$ and satisfy the angular momentum eigenvalue equations

$$\hat{l}^2Y_{lm}(\vartheta,\varphi)=\hbar^2l(l+1)Y_{lm}(\vartheta,\varphi),\quad l=0,1,2,\ldots, \tag{4.9}$$

$$\hat{l}_zY_{lm}(\vartheta,\varphi)=\hbar mY_{lm}(\vartheta,\varphi),\quad m=-l,-l+1,\ldots,0,l+1,\ldots,+l, \tag{4.10}$$

defining the quantum numbers l of orbital momentum and their projection m on the z axis. After substituting (4.8) into (4.5), we obtain the radial Schrödinger equation

$$-\frac{\hbar^2}{2m}\frac{d^2u(r)}{dr^2}+\frac{\hbar^2}{2m}\frac{l(l+1)}{r^2}u(r)+V(r)u(r)=Eu(r),$$

$$V(r)=-\frac{Ze^2}{r}, \tag{4.11}$$

which can be solved for hydrogen atoms or ions having one electron. The solution of Eq. (4.11) can be found analytically using asymptotical solutions proving that an electron cannot get into the center of an atom or move far away

$$\lim_{r \to 0} \cdot u(r) = u_0(r) = r^{l+1},$$

$$\lim_{r \to \infty} \cdot u(r) = u_\infty(r) = e^{-kr}, \quad -k^2 = \frac{2mE}{\hbar^2}.$$

(4.12)

Then the wave function for bound states of an electron (4.11) can be found using the following expression:

$$u_{nl}(r) = e^{-kr} r^{l+1} \sum_{v=0}^{\infty} b_v r^v, \quad b_{v+1} = \frac{2k(v+l+1) + CA}{(v+1)(v+2l+2)} b_v,$$

$$b_0 = 1, \quad C = 2m/\hbar^2, \quad A = -Ze^2.$$

(4.13)

Interrupting series (4.13) on $v = N$, $N = 0, 1, 2, \ldots$, and requiring the finiteness of the solution, we obtain

$$k_n = -\frac{Ze^2}{2n}, \quad n = N + l + 1, \quad E_n = -\frac{mZ^2e^4}{2\hbar^2n^2} = -\frac{Z^2e^4}{(4\pi\varepsilon_0)^2} \frac{m}{2\hbar^2n^2},$$

(4.14)

where the eigenvalues of energy E_n are for the principal quantum numbers n. From the first expression of E_n and radii of quantum diffusion of electrons A[21], we obtain Bohr radii [1] as follows:

$$R_n = \frac{\hbar^2 n^2}{mZe^2}, \quad R_1 = \frac{\hbar^2}{me^2} = 0.529167 \times 10^{-8} \text{ cm.}$$

(4.15)

Radial eigenfunctions u_{nl} are degenerated, and for the same energy E_n, we have the orbital quantum numbers

$$l = 0, 1, 2, \ldots, n - 1,$$

(4.16)

which can be defined by the alphabets $s, p, d, f, g, h, \ldots$. The quantum numbers n, l, m define the eigenvalues for quantum operators $H, \hat{l}^2, l_z$ and wave functions

$$\psi(r, \vartheta, \varphi) = e^{-kr} r^{l+1} \sum_{i=0}^{n-l-1} a_i r^i Y_{lm}(\vartheta, \varphi),$$

(4.17)

having $N = n - l - 1$ knots or cuts of r variable for the radial wave function (4.13) $u_{nl}(r)$. The spherical harmonics form an orthonormal set

$$\int_0^{2\pi} \int_0^{\pi} Y_{lm}^*(\vartheta, \varphi) Y_{l'm'}(\vartheta, \varphi) \sin \vartheta d\vartheta d\varphi = \delta_{ll'} \delta_{mm'}, \qquad (4.18)$$

and under coordinate $\vec{r}$ reflection, $-\vec{r}$ has the parity $(-1)^l$. In the atoms, not only orbital moments but also intrinsic moments of electrons or spins and their magnetic moments are important. For the electromagnetic forces of the external magnetic field B_z, directed in the z direction, which are acting on atoms with magnetic orbital moments μ_l, created by electrons moving in the orbital in the atoms, and magnetic moments μ_s of electrons connected with spin, we obtain

$$F_z = \frac{d}{dz}(\mu_z B_z), \quad \mu_z = \mu_l + \mu_s, \quad \mu_l = -\frac{\mu_0}{\hbar} L_z - \frac{\mu_0}{\hbar} 2S_z, \quad \mu_0 = \frac{e\hbar}{2mc}. \tag{4.19}$$

For the Bohr magneton μ_0, for the single continuous trace of a beam of atoms, when $L_z = \hbar m_l = 0$, the splitting was defined by S_z values. In the Stern–Gerlach experiments, deflection of the trace of silver atoms was measured, having one external electron with the spin moment quantum number $s = 1/2$ and z component of the angular momentum $s_z = \pm \frac{1}{2}\hbar$. By measuring the splitting of the beam in two traces for spin up and spin down with quantum number $s = 1/2$, projection $m_s = \pm \frac{1}{2}$ was obtained. Eigenfunctions of the spin function $\chi_{sm_s}(\psi)$ for the commuting spin operators have eigenvalues

$$\hat{S}^2 \chi_{sm_s} = \hbar^2 s(s+1) \chi_{sm_s}, \tag{4.20}$$

$$\hat{S}_z \chi_{sm_s} = \hbar m_s \chi_{sm_s} \tag{4.21}$$

and depend on spin coordinates ψ having two values $1/2$ or $-1/2$. The function $\chi_{sm_s}(\psi)$ differs from zero only when $m_s = \psi$, and the operator $\hat{S}_z$ can be expressed by the Pauli spin matrix [1] σ_z component as follows:

$$\hat{S}_z = \frac{\hbar}{2} \left\| \begin{matrix} 1 + 0 \\ 0 - 1 \end{matrix} \right\| = \frac{\hbar}{2}\sigma_z. \tag{4.22}$$

Other components of the Pauli matrix $\vec{\sigma}$ can be obtained by requiring commutation relations [1] that are similar in those for orbital angular momentum components. Including two possible projections of spins and quantum numbers n, l and m_l, we can find the number of different quantum states of an electron in the hydrogen atom as follows:

$$2 \sum_{l=0}^{n-1} (2l + 1) = 2n^2. \tag{4.23}$$

We use the fact that for central Coulomb potential the energy values (4.14) do not depend on l or are degenerate. If the potential and force field are only central but not Coulomb, the energy of the particles depends on the quantum numbers n and l. There is a similar situation when we have atoms with more electrons and the Coulomb repulsive forces between them must be included. In this case, considering the Pauli's exclusion principle, we infer that the shell with definite energy can have a maximum number of electrons $2(2l+1)$. This quantum theory explains the periodic table of atoms and their distribution in seven periods with different quantum numbers n and eight groups with different quantum numbers n and l. The first period begins with a hydrogen atom and finishes with helium. The periods that follow begin with the first vertical group of atoms of alkali metals with one electron in the external subshell (Li, Na, K, Rb, Cs, Fr) and finish with inertial atoms (He, Ne, Ar, Kr, Xe, Rn). To illustrate, we present some electronic configurations for metals of the first group Li $1s^2 2s$, Na $1s^2 2p^6 3s$, K $1s^2 2p^6 3s^2 3p^6 4s$ and the inertial eighth group He $1s^2$, Ne $1s^2 2s^2 2p^6$, Ar $1s^2 2s^2 2p^6 3s^2 3p^6$. Going down in groups we have the similar outer shell structure but different number of subshells and similar chemical properties. In the second group of alkali metals, we have the external subshell of electrons ns^2. The eighth or zero group of inertial atoms have a low chemical activity because all possible states in the external subshell or excited physical vacuum are filled with electrons satisfying the Pauli's exclusion principle. In the middle groups, we have s, p, d, f blocks of transition elements with the increasing number of electrons in outer shells. At the bottom of the groups, we have metals. For an increasing number of electrons in the period, we have the change from electropositive elements to non-metallic electronegative elements in chemical reactions. Non-metallic elements are placed toward the right and the top of the table. The last two stable elements are $^A_{82}$Pb, $A = 206, 207, 208$ and $^{209}_{83}$Bi. For large stable nuclei, the number of neutrons is greater than the number of protons Z. In Dubna, Russia, investigators studied the collision of nuclei in the reaction ^{244}Pu$+^{48}$Ca for producing super-heavy elements. One element, $^{289}_{114}$Fl, which was not stable, was found to have a half-life of about 1 minute [2]. For the nucleus $^{298}_{114}$Fl, we have the magic number of neutrons $A - Z = 184$, according to the shell model and the calculations and by using the semi-relativistic model [3], we conclude that this nucleus can be stable [4], A[17]. Using calculations by including the relativistic corrections, the stable nuclei A[16] $^{328}_{114}$X, $^{334}_{120}$X, $^{340}_{126}$X with magic number protons and neutrons were obtained. Contributions to the magnetic moment of atom (or nucleus) consist of the orbital motion of electrons (or protons) and intrinsic spin of electrons (neutrons). The interaction forces of the total angular moment $\vec{J}$ can be calculated by using two different coupling methods LS coupling:

$$\vec{L} = \sum_{i=1}^{N} \vec{l}_i, \quad \vec{S} = \sum_{i=1}^{N} \vec{s}_i, \quad \vec{J} = \vec{L} + \vec{S}, \tag{4.24}$$

and JJ coupling

$$\vec{j}_i = \vec{l}_i + \vec{s}_i, \quad J = \sum_{i=1}^{N} j_i. \tag{4.25}$$

In the central field with an additional spin–orbit interaction, the energies of the electron (or nucleon's) in the shells are characterized by the orbital quantum numbers l and the total quantum number j

$$j = l \pm \frac{1}{2}. \tag{4.26}$$

When the interaction between the magnetic moments of electrons and the orbital magnetic moments is weak, usually for atoms we have SL coupling and energies and the quantum states of atoms are defined by the terms ^{2S+1}L, $L = 0, 1, 2, \ldots = S, P, D, \ldots$. The total angular quantum numbers J are changing in the interval $J + S, J + S - 1, \ldots, |J - S|$. Hence, the simultaneously defined eigenfunctions depend on the commuting operators $\vec{J}^2, J_z, \vec{L}^2, \vec{S}^2$. The spinning electrons interacting with the central Coulomb field of the atom add the following term in the Hamiltonian of weak interaction:

$$\Delta E = W(r)(\vec{L}\vec{\sigma}) = W(r)\frac{1}{\hbar}(\vec{L}\vec{S}), \qquad (4.27)$$

and generate small changes of one-particle energy in the first approach [1]

$$\Delta E = \left\{ \begin{matrix} l\hbar \\ -(l+1)\hbar \end{matrix} \right\} \int_0^\infty u_{nl}^*(r)W(r)u_{nl}(r)dr \quad \text{for} \quad \left\{ \begin{matrix} j = l + 1/2 \\ j = l - 1/2 \end{matrix} \right\}, \qquad (4.28)$$

by using the following equation:

$$\vec{s} \cdot \vec{l} = \frac{1}{2}(\vec{j}^2 - \vec{l}^2 - \vec{s}^2). \qquad (4.29)$$

The spin–orbit interaction [1] in atoms can be expressed as

$$H_{SL} = \frac{1}{2m^2c^2}\frac{1}{r}\frac{dV}{dr}\vec{l}\vec{s}, \qquad (4.30)$$

and is similar to nucleons in nuclei [5]

$$H_{sl} = -\lambda\frac{1}{m^2c^2}\frac{1}{r}\frac{dV}{dr}\vec{l}\vec{s}, \quad \lambda \approx 10, \qquad (4.31)$$

where λ can be defined by bound state energies of nuclei or experiments on the scattering of nucleons.

In this case, wave functions can be represented using Clebsch–Gordan coefficients [1]:

$$\psi_{lsjm} = u_{nl}(r)Y_{lsjm}$$

$$= u_{nl}(r)\sum_m (lm_l sm_s|jm)Y_{lml}(\vartheta, \varphi)\chi_{sm_s}, \quad m = m_l + m_s.$$

$$(4.32)$$

For the quantum one-particle states, when spin–orbit interactions are included, the sub-shells can be represented as follows:

$$1s_{1/2} \quad 2s_{1/2} \quad 2p_{1/2} \quad 2p_{3/2} \quad 3s_{1/2} \quad 3p_{1/2} \quad 3p_{3/2} \quad 3d_{3/2} \quad 3d_{5/2},$$

$$(4.33)$$

and the shells for atoms are represented by $K, L, M, N, O, P, Q, \ldots$ with principal quantum numbers

$$n = 1, 2, 3, 4, 5, 6, \ldots \tag{4.34}$$

For example, for subshells, we note that $L_1 \to 1s_{1/2}$, $L_2 \to 2s_{1/2}$, $L_3 \to 2p_{3/2}$.

A strong spin–orbit term (4.31) in the nucleus generates large degeneracy of orbitals and gives an idea of the structure of nucleons, with the shells producing stable nucleon structures for magic numbers [5] N, $Z = 2, 8, 20, 28, 50, 82, \ldots$. For the Hamiltonian (4.2), where interactions between particles V_{jk} are included, the wave function satisfying the Pauli's exclusion principle for two Fermi particles with spins $s = \frac{1}{2}$ can be expressed by one-particle wave functions $\varphi(x)$ like its determinant

$$\varphi_a^0(x_1, x_2) = \frac{1}{\sqrt{N!}} \begin{vmatrix} \varphi_1(x_1)\varphi_1(x_2) \\ \varphi_2(x_1)\varphi_2(x_2) \end{vmatrix}. \tag{4.35}$$

References

1. Merzbacher, E. (1970). *Quantum Mechanics*, John Wiley & Sons, New York.
2. Oganessian, Yu. Ts. *et al.* (2000). The synthesis of superheavy nuclei in the ^{244}Pu $+^{48}$Ca reaction, *Nucl. Phys.*, **63**(10), 1769–1777 (Russia).
3. Janavičius, A. J. (1996). Relativistic corrections in the average field of neutron and proton shells, *Acta Phys. Pol. B*, **27**, 2195–2205.
4. Janavičius, A. J. (2006). Mathematical methods in the semi-relativistic single-particle model for superheavy nuclei, in *Proceedings of the 10th World Multi-Conference on Systemics, Cybernettics and Informatics*, Vol. 3, IIIS, Orlando, pp. 66–68.
5. Heyde, K. L. G. (1994). *The Nuclear Shell Model*, Springer-Verlag, New York.

Chapter 5

The Scattering Theory
and Nuclear Reactions

5.1 Introduction

In this chapter, new mathematical methods and physical models
are presented and applied in Refs. A[1, 2] scattering theory for
studying the structure of a nucleus and its interactions with the
elementary particles. The theoretical investigation of the scattering
of two particles from the center of masses can be carried out by
using the potential representation method A[1, 2]. In the simplest
cases, the solutions of the Schrödinger equations for the probabilities
of particle scattering can be obtained analytically for investigating
the interaction potentials $V(r)$ between nuclei. Realistic potentials
among protons, neutrons, nucleus and other particles are complicated
and consist of strong Coulomb interactions. The most interesting
scattering experiments can be realized for non-relativistic particles
with energies less than $10\,\mathrm{MeV}$. In nuclear physics, energy is
usually expressed by electron volts $1\,\mathrm{eV} = 1.602 \cdot 10^{-19}\,\mathrm{J}$. In the
following chapters, the new perturbation method, named potential
representation, is presented and considered for evaluating the positive
energies and bound states in one-particle and many-particle cases.
Jost functions that directly depend on Woods–Saxon potential are
considered A[1]. The possibility to express the wave function like a
product of the unperturbed solution and the function which depends
on perturbation potential provide a wide scope for modeling in all

branches of quantum mechanics and atomic, molecular, and nuclear physics.

The majority of what we know about the forces in the nuclei and between two nuclei or particles has been investigated using scattering experiments. The particles used in the beams for bombardment, usually used as projectiles, are neutrons of nuclei, protons, alpha particles and ions. These particles are scattered by the nuclei and the measurements of intensity of the scattered particles depend on the scattering angle and their energy thus giving information about the structure of the nuclei, nucleon density, charge distribution and the interaction forces. We must choose kinetic energies E_{kin} for interaction potentials $V(r)$ requiring $E_{\text{kin}} - V(r) \geq 0$.

Then, for positive eigenvalues of the Schrödinger equation, we obtain the eigenfunctions of unbound states. The relation between the intensities of the incident particles with the calculated wave functions defining the dependence of intensities of distribution from the scattering angles of particles elastically scattered from the target nuclei is the main topic of study of the scattering theory. The scattering of particles depends on the energy E of the particles and interaction potential and can be defined by the Schrödinger equation

$$-\frac{\hbar}{2\mu}\nabla^2\psi + V(r)\psi = E\psi. \tag{5.1}$$

The solution $\psi(r)$ is called the distorted wave, and it describes the scattering of the particles and their absorption, when imaginary potential $iW(r)$ is included, i.e.,

$$U(r) = V(r) + iW(r). \tag{5.2}$$

The simplest case is the scattering of a monoenergetic low-density beam of particles moving toward the scatterer from a large distance. In this case, the particles are interacting only with particles in the scatterer. We will specialize in the scattering theory without energy loss or elastic scattering. The scattering probability of particles scattered in a cone of a solid angle $d\Omega$ at the origin is defined by the differential scattering cross-section

$$d\sigma = \frac{I(\vartheta, \varphi)}{I_0}d\Omega, \tag{5.3}$$

where I_0 is the number of particles in the incident beam moving from the left per unit area and $Id\Omega$ is the number of particles scattered into the cone. $\sigma(\vartheta, \varphi)$ can be found by solving the Schrödinger equation for relative movement in the coordinates of the center of mass system for relative energy E and reduced mass μ

$$E = \frac{m_2}{m_1 + m_2} E_{\text{kin}}, \quad \mu = \frac{m_1 m_2}{m_1 + m_2}. \tag{5.4}$$

For $m_1 = m_2 = m$, we have $E = E_{\text{kin}}/2$, half of kinetic energy E_{kin} in laboratorial coordinates, and $\mu = m/2$. The scattering of particles with reduced masses μ are defined by the asymptote which consists of wave or particle spreading with velocity v in the z direction and wave spreading from the scattering center [1, 2]

$$\lim_{r \to \infty} \psi(r, \vartheta, \varphi) = A[e^{ikz} + r^{-1}f(\vartheta, \varphi)e^{ikr}], \quad k = \sqrt{2mE}/\hbar, \tag{5.5}$$

where the first term corresponds to the incident beam moving in the z direction and the second term is the scattered wave. By satisfying the condition $\frac{\partial}{\partial t}\int \psi^*\psi dV = 0$, we can see that [3] the flow of scattered particles is given as

$$v|A|^2|f(\vartheta, \varphi)|^2/r^2 \tag{5.6}$$

The differential scattering cross-section was thus obtained as

$$\sigma(\vartheta, \varphi) = |f(\vartheta, \varphi)|^2 d\Omega. \tag{5.7}$$

Constant A can be chosen by requiring that falling flow density would be equal to unity $A = 1/\sqrt{v}$, and for other cases, $A = 1$.

Solution for a spherically symmetric interaction potential [3] of (5.1)

$$\psi = \sum_{l=0}^{\infty} A_l R_l(r) P_l(\cos \vartheta), \quad A_l = (2l + 1)i^l e^{i\delta_l} \tag{5.8}$$

satisfies the following asymptotic expression:

$$\lim_{r \to \infty} R_l(r) \cong \frac{1}{kr} \sin\left(kr - \frac{l\pi}{2} + \delta_i\right), \quad kr \gg l, \tag{5.9}$$

defining the phase shifts δ_l and wave function

$$\psi = \sum_{l=0}^{\infty}(2l+1)i^l e^{i\delta_l} R_l(kr)P_l(\cos\vartheta). \qquad (5.10)$$

From the last formula and (5.5), the equation for a scattered wave

$$\psi_{SC} = f(\vartheta)\frac{e^{ikr}}{r} \qquad (5.11)$$

can be obtained as follows by defining the number [3] dn of scattered particles, moving with velocity v in the volume angle $d\Omega$ after reflection from the surface dS:

$$dn = \left|f(\vartheta)\frac{e^{ikr}}{r}\right|^2 vdS = |f(\vartheta)|^2\frac{v}{r^2}dS = |f(\vartheta)|^2 vd\Omega. \qquad (5.12)$$

Once we know the value of $dn(\vartheta)$, we can find the expression of the scattering differential cross-section [3]

$$\frac{d\sigma}{d\vartheta} = |f(\vartheta)|^2, \qquad (5.13)$$

and the total scattering cross-section [3]

$$\sigma_t = \int_0^{2\pi} d\varphi \int_0^{\pi} |f(\vartheta)|^2 \sin\vartheta d\vartheta = \frac{\pi}{k^2}\sum_{l=0}^{\infty}(2l+1)|1-S_l|^2, \qquad (5.14)$$

from the scattering amplitude. The experimentally verified absorption can be defined from the relative decreasing flow [3]

$$\frac{I(\Delta z)}{I(0)} = \exp[-\sigma_t \cdot n \cdot \Delta z], \qquad (5.15)$$

after passing through the target with the nuclear density n and target width Δz.

The scattering amplitude [3] for cylindrically symmetric case can be expressed by phase shift $\delta_l(k)$

$$f_k(\vartheta) = \frac{1}{k}\sum_{l=0}^{\infty}(2l+1)e^{i\delta_l}\sin\delta_l P_l(\cos\vartheta), \qquad (5.16)$$

and scattering matrix

$$S_l = e^{2i\delta l}. \qquad (5.17)$$

The differential scattering cross-sections

$$\frac{d\sigma}{d\Omega} = |f_k(\vartheta)|^2 = \frac{1}{k^2}\left|\sum_{l=0}^{\infty}(2l+1)e^{i\delta_l}\sin\delta_l P_l(\cos(\vartheta))\right|^2, \qquad (5.18)$$

and total cross-sections [3]

$$\sigma_t = \int \frac{d\sigma}{d\Omega}d\Omega = \frac{4\pi}{k^2}\sum_{l=0}^{\infty}(2l+1)\sin^2\delta_l, \qquad (5.19)$$

for scattering amplitude having a cylindrically symmetric case which are expressed by the phase shifts $\delta_l(k)$ and scattering matrix $e^{2i\delta_l}$ [3]

$$f_k(\vartheta) \cong \frac{1}{2ik}\sum_{l=0}^{\infty}(2l+1)(1-e^{2i\delta_l})P_i(\cos\vartheta). \qquad (5.20)$$

The scattering of the particles for different quantum numbers of orbital angular momentum l depends on energy and potential $V(r)$. The partial wave amplitudes are defined for the decreasing potential at large distances and are strongest at origin. Then, the only low-angular momentum components of the scattering amplitude have significant values. The phase shift δ_l for the impact parameter ρ for the scattered particle is strongest at the origin and decreases at large distances [3]

$$\rho = \frac{l\hbar}{p} = \frac{l}{k}, \qquad (5.21)$$

and tends to zero for the potential with radius a for orbital quantum number when l exceeds ka. The elastic scattering matrix $S_l = e^{2i\delta_l}$ has real δ_l and also $S_l^* \cdot S_l = 1$. For non-elastic processes, $S_l^* \cdot S_l < 1$ phase shifts are taken as complex values, $\delta_l = e + ir$. The complex potential representing absorption generates the following reaction or absorption cross section [3]:

$$\sigma_{rl} = \frac{\pi}{k^2}(2l+1)(1-|S_l|^2). \qquad (5.22)$$

The total cross-section consisting of scattering and reaction cross-sections can be defined [3] as follows:

$$\sigma_{tl} = \frac{\pi}{k^2}(2l+1)(1-\operatorname{Re}S_l), \qquad \lim_{k\to 0}\sigma_{t0} = 4\pi a^2 \qquad (5.23)$$

by the real part of the scattering matrix. For the limit of zero energy σ_l, defined by an impenetrable sphere of radius, a is usually used for projectiles like neutrons. When the impact parameter exceeds the range a of the potential, when $l > ka$, no scattering occurs and in (5.18), (5.19) we have a small number of quantities in sums because the vanishing phase shifts at large l quantum numbers equal to zero.

5.2 Nuclear Reactions and the Optical Model

The first nuclear reaction of scattering α particles from the radioactive nuclei $^{214}_{84}$Po was observed by Ernest Rutherford (1919), using its initial kinetic energies E_l (7.683 MeV) in gas $^{14}_{7}$N under fixed pressure. The transmutation of elements was obtained by the following reaction [4]:

$$^4_2\text{He} + E_1 + {}^{14}_{7}\text{N} \rightarrow {}^{17}_{8}\text{O} + {}^1_1\text{H} + E_2, \tag{5.24}$$

which can be written in short as follows:

$$^{14}_{7}\text{N}(^4_2\text{He}, {}^1_1\text{H})^{17}_{8}\text{O}. \tag{5.25}$$

Here, with energies E_2 (6.5 MeV) fast protons $^1_1$H having a range 0.40 m in air at normal condition were detected. It is significantly greater than 0.07 m for the alpha particles $^4_2$He spreading with strong ionizing and initial energy $E_2 = 6.5$ MeV.

Many nuclear reactions occur in two separate stages after the formation of the compound nucleus with a half-life $\tau_{1/2} \approx 10^{-16}$ s. These reactions consist of the target nucleus, incident particles and form compound excited nuclei. We present some reactions for compound nuclei

$$^{14}_{7}\text{N} + {}^1_0n \rightarrow {}^{15}_{7}\text{N}^*(0.5\,\text{MeV}), \quad {}^{14}_{7}\text{N} + {}^1_1\text{H} \rightarrow {}^{15}_{8}\text{O}^*(7.5\,\text{MeV}), \tag{5.26}$$

For incident particles with high energies, we have direct, very fast reactions over nuclear time

$$\tau_{nt} \approx 10^{-21}\text{s}. \tag{5.27}$$

The decay of the compound nucleus may be de-excited by one or many different ways depending on its energy of excitation. In this case, we obtain the recoil nucleus and ejected particle. The different decay modes [4] are obtained as presented below

$$^{14}_{7}N^*(12\,\text{MeV}) \rightarrow \, ^{13}_{7}N + \, ^{1}_{0}n, \quad ^{14}_{7}N^*(12\,\text{MeV}) \rightarrow \, ^{13}_{6}C + \, ^{1}_{1}H,$$

$$^{14}_{7}N^*(12\,\text{MeV}) \rightarrow \, ^{12}_{6}C + \, ^{2}_{1}H, \tag{5.28}$$

and by other additional channels. The compound nucleus modes can be successfully applied for explaining the nuclear reactions for incident particles with more than 15-MeV energy and target $A > 10$.

The compound nuclei that are presented are de-excited by ejecting small particles and produce new nuclei. For these long-lived states (10^{-16}s), the uncertainties of compound nuclei energies are about $6.6\,\text{eV}$ and virtual states with positive energies are generated by quantum diffusion A[21]. The compound nucleus reactions can be presented [4] by the following relations:

$$a + X \rightarrow C^* \rightarrow Y + b, \quad \sigma_{ab} = \sigma_a \Gamma_b / \hbar,$$

$$\omega_b = 1/\tau_b = \Gamma_b / \hbar, \tag{5.29}$$

where ω_b is the total decay probability, Γ_b is the partial width and σ_a is the cross-section of the compound nucleus formation.

The direct reactions with interaction time $10^{-22}s$ do not produce intermediate states and the reaction cross-sections are spreading in the region of energies of about $6.6\,\text{MeV}$. For elastic scattering at great distances, we have the superposition of falling e^{-ikr}/r and scattered e^{ikr}/r waves with equal amplitudes. For radioactive fission of nucleus, we have only the spreading wave

$$\psi(r,t) = \psi(r) \cdot e^{-\frac{iEt}{\hbar} - \frac{wt}{\hbar}} = \psi(r) \cdot e^{-\frac{i}{\hbar} \cdot [E - i\frac{\Gamma}{2}] \cdot t} \tag{5.30}$$

which is compared with the well-known macroscopic formula of the radioactive decay law

$$\frac{dN(t)}{dt} = -wN, \quad N = N_0 e^{-wt} = \text{const} \int \psi * (r,t) \psi(r,t) d\vec{r}, \tag{5.31}$$

where N is the number of radioactive nuclei. The probability of fission is obtained by introducing the complex energy $E - i\Gamma/2$ and comparing (5.30) and (5.31),

$$w = \Gamma/\hbar \qquad (5.32)$$

using the above equation, the average or mean lifetime τ of a single atom and the half-life $\tau_{1/2}$ can be expressed as follows:

$$\tau = \frac{1}{w}, \quad \tau_{1/2} = \frac{\ln 2}{w} = \frac{0.693}{w} = \tau \ln 2, \qquad (5.33)$$

when half of parent nuclides stay undecayed.

The use of the optical potential for particles' scattering with absorption can be realized by introducing the complex potential $V + iW$ in the Schrödinger equation. For considering diffraction and scattering phenomena with the optical model, we must use the Woods–Saxon potential of the shell model. We added an imaginary part of about few MeV, which is consistent [4] with the lifetime $10^{-22}s$ of a single particle in the virtual states.

According to the optical model, the nucleus of atoms represents a compact environment which consists of positive protons and neutral neutrons bounded by short radius $2 \cdot 10^{-13}$ cm of attractive forces which at $0.4 \cdot 10^{-13}$ cm become repulsive. The interaction potential $V(r)$ between particles of projectile with protons and neutrons in nuclei is very complicated and we can introduce the average complex potentials (5.2) with the real part [3]

$$V(r) = \int \rho(r_i) V_i(r - r_i) d^3(r - r_i), \qquad (5.34)$$

which can be expressed by the density of the nucleon $\rho(r_i)$ of the target nucleus and the projectile and interaction potential of the nucleons $V_i(r - r_i)$ using the integration of overall nucleus volume. For a composite incident particle, we must include additional density of this particle, interaction potential between all nucleons and double folding. The imaginary part of the optical potential is defined from the absorption of incident particles' flux. The Coulomb potential must also be included in $V(r)$. For a complex potential, the wave vector is also complex, i.e.,

$$k = k_r + ik_m. \qquad (5.35)$$

Then we have the wave function

$$\psi = \exp(ikr) = \exp(ik_r r)\exp(-k_m r), \qquad (5.36)$$

with the following probability density dependence on the distance

$$\psi^*\psi = \exp(-2k_m r). \qquad (5.37)$$

The density attenuation is characterized by a mean free path defining the probability interaction of the incident particle with nucleus

$$\lambda_m = \frac{1}{2k_m}. \qquad (5.38)$$

The wave vector satisfies [5] the equation

$$\frac{\hbar^2 k^2}{2m} = U + E, \quad U(r) = V(r) + iW(r), \qquad (5.39)$$

from which we obtain the following equations:

$$k_r^2 - k_m^2 = \frac{2m}{\hbar^2}(V+E), \quad k_r k_m = \frac{m}{\hbar^2}W. \qquad (5.40)$$

The above equation describes the scattering and absorbed particles for low-energy $E \ll V$ neutrons. The imaginary part of optical potential is proportional to the absorption coefficient (5.22) of the incident particles. For nucleon and heavy ion scattering at low energies, absorption occurs at the nuclear surface. For W and $E \ll V$, using a complex potential enables to describe the scattering of strongly and weakly absorbed particles thus, [5] we have

$$k_m \approx \frac{W}{\hbar}\sqrt{\frac{m}{2V}}. \qquad (5.41)$$

A very common analytical form is the optical potential of the Woods–Saxon potential, i.e.,

$$U(r) = -Vf(r,R,a) - iWf(r,R_m,a_m), \qquad (5.42)$$

$$f(r,R,a) = (e^x + 1)^{-1}, \quad x = (r-R)/a. \qquad (5.43)$$

that has a form similar to the density of the nucleus matter.

The parameters of complex $U(r)$ and real $V(r)$ potentials can be determined by varying the parameters for obtaining the best fit to the scattering matrix A[2]:

$$S_l(k) = \frac{c \int_\infty^0 \frac{dz}{f_0^2(l,k,z)} \int_\infty^z V(y)\varphi(l,k,y) f_0^2(l,k,y) dy + 1}{c \int_\infty^0 \frac{dz}{f_0^2(l,-k,z)} \int_\infty^z V(y)\varphi(l,-k,y) f_0^2(l,-k,y) dy + 1},$$

$$c = \frac{2m}{\hbar^2}, \quad \mu = \frac{m_1 m_2}{m_1 + m_2}. \tag{5.44}$$

This original formula for $S_l(k)$ was obtained using potential representation method [5], where the Jost solutions [6] of the Schrödinger equation for positive energies are represented like products

$$f(l, \mp k, r) = \varphi(l, \mp k, r) \cdot f_0(l, \mp k, r),$$
$$\lim_{r \to \infty} f(l, \mp k, r) \exp(\mp ikr) = 1, \tag{5.45}$$

of free solutions $f_0(l, \mp k, r)$ on multipliers $\varphi(l, \mp k, r)$, which depend on the interaction potential. It can be found by solving the integral equations A[2]

$$\varphi(l, \mp k, r) = c \int_\infty^r \frac{1}{f_0^2(l, \mp k, y)} dy \int_\infty^y V f_0^2(l, \mp k, x) \varphi(l, \mp k, x) dx + 1, \tag{5.46}$$

for $\varphi(l, k, r)$ and $\varphi(l, -k, r)$. In this case, we must use the free Jost solutions expressed by Hankel functions [6]

$$f_0(l, k, x) = \left(\frac{\pi}{2}\right)^{1/2} \exp\left\{-i\frac{\pi}{2}(l+1)\right\} (kx)^{1/2} H_{l+\frac{1}{2}}^{(2)}(kx), \tag{5.47}$$

$$f_0(l, -k, x) = \left(\frac{\pi}{2}\right)^{1/2} \exp\left\{i\frac{\pi}{2}(l+1)\right\} (kx)^{1/2} H_{l+\frac{1}{2}}^{(1)}(kx). \tag{5.48}$$

Substituting like first approach $\varphi_0 = 1$ in integral (5.46), after integration we can find φ_1 and then repeat the procedure till desirable coincidence is achieved. Then we can find [6] the scattering matrix

(5.44)

$$S_l(k) = \exp[2i\delta_i] \frac{(2l+1)\lim\limits_{r\to 0} r^l f(l,k,r)}{(2l+1)\lim\limits_{r\to 0} r^l f(l,-k,r)} = \frac{f(l,k)}{f(l,-k)} \exp[i\pi \cdot l],$$

(5.49)

expressed by the Jost functions $f(l,k)$ and $f(l,-k)$. Standard, more complicated methods of scattering wave phase shift calculations are based on comparing the solutions ψ of (5.1) and $\frac{d\psi}{dr}$ on the surface of the nucleus R and the asymptotical solutions (5.9) for the positive scattering energies $E > 0$ of the particles. From the scattering experiments, we can find the differential (5.12), total (5.14) and reaction (5.22) cross-sections and calculate the equations (5.44) and (5.46) using model potentials. Potentials cannot be defined uniquely by following this method. For low-energy nucleon scattering, the typical values of the real V and complex W [4] are 50 MeV and 3 MeV respectively. From (5.20) and (5.22), when $E \ll V$ was considered, the mean free path was determined to be $\lambda_m = 10$ fm $(10^{-12}$ cm$)$. The excitation energy in the target nucleus consists of incident nucleon energy and the bound energy of the nucleons $(-8$ MeV$)$. The great free path of nucleons is in good agreement with the independent-particle model of the nucleus. The radii of nuclei can be obtained [4] by keeping in mind its constant density

$$R = r_0 \cdot A^{1/3}, \quad r_0 \approx 1.2 \text{ fm}, \tag{5.50}$$

where A is the mass number or number of nucleons. The polarization effects of scattering in optical model and the shell structure of nuclei can be explained by introducing the spin–orbit potentials

$$V_{sL}(r) = -V_{s0}\left(\frac{\hbar}{M_\pi c}\right)^2 \frac{1}{r}\frac{df(r)}{dr}(\vec{l}\cdot\vec{s}), \tag{5.51}$$

where $f(r)$ is the function type of Woods–Saxon potential and $(\hbar/M_\pi c)^2 \approx 2$ fm^2. The expectation values $(\vec{l}\cdot\vec{s})$ can be obtained by using the relation for the square of total angular momentum,

$$\vec{j}^2 = (\vec{l} + \vec{s})^2$$

$$(\vec{l} \cdot \vec{s}) = \frac{1}{2}(j(j+1)) - l(l+1) - s(s+1)),$$

$$= \frac{1}{2}l, \quad j = l + \frac{1}{2},$$

$$= -\frac{1}{2}(l+1), \quad j = l - \frac{1}{2}. \tag{5.52}$$

The splitting of the spin–orbit interaction potential depends on j equal sum of orbital l and spin quantum numbers. The wave functions, where spin–orbit interactions are included for positive and negative energies, can be represented using the following series of Clebsch–Gordan coefficients:

$$\psi_{nljm}(r, \vartheta, \varphi) = \frac{u_{nlj}(r)}{r} Y_{ljm}(\vartheta, \varphi), \tag{5.53}$$

$$\psi_{nljm}(r, \vartheta, \varphi) = \sum_{m_l, m_s} \langle lm_l sm_s | jm \rangle Y_{lm_l}(\vartheta, \varphi) \chi_s(m_s), \tag{5.54}$$

$$s = 1/2, \quad m = m_1 + m_s, \quad j = 1 \pm 1/2,$$

$$l = 0, 1, 2, \ldots; \quad s, p, d, e, f, \ldots. \tag{5.55}$$

In this case, the quantum states of the nucleon are represented as nl_j, where $n = 1, 2, 3, \ldots$ is the principal quantum number and $j = l + 1/2$ or $j = l - 1/2$, $j = 1/2; 3/2; 5/2, \ldots$, is the momentum quantum number and l is the orbital angular momentum quantum number.

5.3   Inverse Tasks of Scattering

Some of the more complicated tasks are finding the interaction potentials from experimentally measured differential (5.12) and total (5.15) cross-sections, which must agree with the scattering matrix. The aim of finding the potentials can be realized by analytical properties of the scattering matrix in the complex orbital momentum plain. In the paper [9], the method for finding l independent α–α scattering model potential was presented:

$$V(r) = V_1 \frac{e^{-\mu_1 r}}{r} + V_2 \frac{e^{-\mu_2 r}}{r}, \tag{5.56}$$

from the poles of scattering matrix at the definite energy using the experimental phase shifts for l_0, l_2. The Woods–Saxon potential

$$V(r) = \frac{V_0}{1 + \exp\frac{r-R}{a}}, \qquad (5.57)$$

defined by depth $V_0 < 0$, width R and surface thickness a at low energies generates particle scattering on the nuclei. Analytical expression for the scattering matrix can be found for low energies using potential representation method. From analytical expression of Jost solutions A[1] of the radial Schrödinger equation,

$$\frac{d^2}{dr^2}\varphi \cdot f_0 + [k^2 - cV(r)]\varphi(r)f_0 = 0,$$
$$\varphi = \varphi(\pm k, V(r)), \qquad f_0 = \exp(\mp kr) \qquad (5.58)$$

at $l = 0$, we obtained A[1] the scattering matrix

$$S(k, l = 0) = \frac{\varphi(+k, V_p)}{\varphi(-k, V_p)}, \qquad V_p = V(r = 0), \qquad (5.59)$$

and its analytical expression

$$S(l = 0) = \exp\left[i2\alpha\left(-\frac{2\beta\ln 2}{1 + 4\alpha^2} + \ln 2 - \frac{R}{a}\right)\right] \cdot Z, \qquad (5.60)$$

$$Z = \frac{ctg\gamma - \dfrac{i\alpha}{\sqrt{\alpha^2 - \beta}} - \dfrac{\beta(1 - 2i\alpha)}{\sqrt{(\alpha^2 - \beta)(1 + 4\alpha^2)}}}{ctg\gamma + \dfrac{i\alpha}{\sqrt{\alpha^2 - \beta}} - \dfrac{\beta(1 + 2i\alpha)}{\sqrt{(\alpha^2 - \beta)(1 + 4\alpha^2)}}},$$

$$\gamma = \sqrt{\alpha^2 - \beta} \cdot (\ln 2 - R/a), \qquad \alpha = ak, \qquad \beta = a^2 V_0 c, \qquad c = \frac{2m}{\hbar^2}. \qquad (5.61)$$

Using potential representation of Jost solutions $f(\mp k, r) = \varphi(\mp k, r) \cdot \exp(\pm ik, r)$, the function φ was expressed like power series of the Woods–Saxon potential,

$$\varphi(\mp k, V(r)) = \sum_{n=0}^{\infty} a_n(\pm k, V) \cdot V^n. \qquad (5.62)$$

For this case, the Jost solutions of the Schrödinger equation (5.58) were used for obtaining (5.61) the analytical expression of the scattering matrix (5.58). The scattering matrices obtained from the experimental results can be used for defining the parameters R, a, V_0 of the Woods–Saxon potential.

For the scattering of protons, alpha particles, on the nuclei, we performed the separation of the nuclear forces and Coulomb interactions, which is a very important task. The potential representation method was used for obtaining the integral equations, thus achieving this separation A[6]. It can be useful for considering and comparing proton–proton, neutron–neutron and neutron–proton scattering and nucleons interactions [4]. The potentials found from measurements of differential cross-sections (5.12) and total cross-sections (5.15) can be compared with theoretically calculated values (5.13), (5.15) from the scattering matrix and phase shifts [6] expressed by Jost functions. These functions can be found by solutions of the integral equations using the method of potential representation A[1, 2]. For positive energies, we presented the integral equations, the solution for which can be obtained by multiplying the known unperturbed free solutions having asymptotes at infinity $e^{\mp ikx}$ on the multiplier, which depends on the perturbation potential. This method can also be used for analytical solutions for the definition of a scattering matrix's dependence on the potential parameters A[2]. In addition, this method can be used for the introduction of relativistic effects in scattering matrix calculations A[4]. We solved the obtained integral equations (5.35) by the iteration method for finding the scattering matrices (5.33) with high accuracy.

References

1. Merzbacher, E. (1970). *Quantum Mechanics*, Wiley, New York, p. 621.
2. Schiff, L. I. (1955). *Quantum Mechanics*, McGraw-Hill Book Company, Inc., New York, p. 473.
3. Wilhelmi, Z. (1976). *Physics of Nuclear Reactions*, State Science Publisher, Warsaw, p. 450 (in Polish).
4. Devanarayanan, S. (2016). *A Text Book on Nuclear Physics*, Lexington, USA, p. 380.
5. Jelley, N. J. (1990). *Fundamentals of Nuclear Physics*, Cambridge University Press, Cambridge, p. 278.
6. de Alfaro, V. and Regge, T. (1965). *Potential Scattering*, North-Holland Publishing Company, Amsterdam, p. 274.

Chapter 6

The Schrödinger Equation in Potential Representation

6.1 Introduction

The wave function usually depends either on the coordinates or on the momenta; in these cases, we have, respectively, the coordinate or momentum representation. In the case where the potential is a single-valued function of the coordinates and vice versa, the variable which determines the wave function can be expressed in terms of the potential A[1]. In this case, the wave function is a function of the interaction potential. Such a representation is called a potential representation A[1]. This allows to consider some problems of quantum mechanics and capabilities of the proposed method. For this purpose, we represent the solution of the Schrödinger equation in the following form:

$$f(l, \pm k, x) = \varphi(l, \pm k, x) f_0(l, \pm k, x), \qquad (6.1)$$

where the solution $f(l, \pm k, x)$ satisfies the boundary condition

$$\lim_{x \to \infty} f(l, \pm k, x) = \exp\{(\mp ikx)\}. \qquad (6.2)$$

Jost solution $f_0(l, \pm k, x)$ is the solution of the Schrödinger equation in the absence of interaction. From (6.1) and (6.2), we obtain the following boundary condition for the function:

$$\lim_{x \to \infty} \varphi(l, \pm k, x) = 1. \qquad (6.3)$$

Substituting (6.1) in a radial Schrödinger equation, we obtain

$$\frac{d^2}{dx^2}\varphi \cdot f_0 + \left(k^2 - \frac{l(l+1)}{x^2} - cV(x)\right)\varphi \cdot f_0 = 0, \quad c = \frac{2M_r}{\hbar^2}. \quad (6.4)$$

Given that $f_0(l, \pm k, x)$ is a free solution, we obtain

$$f_0\frac{d^2\varphi}{dx^2} + 2\frac{df_0}{dx}\frac{d\varphi}{dx} - \varphi \cdot f_0 cV(x) = 0. \quad (6.5)$$

Since the short-range interactions are taken into account using the function $\varphi(l, \pm k, x)$, it is clear that here a radial coordinate can be replaced by a potential

$$\varphi(x) = \varphi(V(x)). \quad (6.6)$$

Assuming that $\varphi(V(x))$ depends only on the potential, Eq. (6.5) can be rewritten as follows:

$$\left(\frac{dV}{dx}\right)^2\frac{d^2\varphi}{dV^2} + \left(\frac{d^2V}{dx^2} + 2\frac{f_0'}{f_0}\frac{dV}{dx}\right)\frac{d\varphi}{dV} - cV(x)\varphi = 0. \quad (6.7)$$

6.2   Solution in the Case of s-Waves

For s-waves, we have the relation (6.2),

$$\frac{f_0'(\pm k, x)}{f_0(\pm k, x)} = \mp ik. \quad (6.8)$$

Then (6.7) simplifies it as follows:

$$\left(\frac{dV}{dx}\right)^2\frac{d^2\varphi}{dV^2} + \left(\frac{d^2V}{dx^2} \mp 2ik\frac{dV}{dx}\right)\frac{d\varphi}{dV} - cV(x)\varphi = 0. \quad (6.9)$$

In the presented equation, the Woods–Saxon potential for the practical applications

$$V(x) = V_0\left(1 + \exp\left\{\frac{x-R}{a}\right\}\right)^{-1}. \quad (6.10)$$

can be used.

We assume that V_0 can take complex values. From (6.10), it is easy to find

$$\frac{dV}{dx} = \frac{V}{a}\left(\frac{V}{V_0} - 1\right),$$

$$\frac{d^2V}{dx^2} = \frac{V}{a^2}\left(\frac{2V}{V_0} - 1\right)\left(\frac{V}{V_0} - 1\right),$$

(6.11)

Substituting (6.11) into (6.9) and making some transformations, we obtain

$$\frac{V}{a^2V_0^2}(V - V_0)^2\frac{d^2\varphi(\pm, k, V)}{dV^2} + \frac{1}{a^2V_0^2}((V - V_0)^2$$

$$+V(V - V_0) \mp 2ikaV_0(V - V_0))\frac{d\varphi(\pm k, V)}{dV} - c\varphi(\mp k, V) = 0.$$

(6.12)

Since the coefficients of Eq. (6.12) depend only on potential V, we can expand the solution $\varphi(\pm k, v)$ by the potential in power series

$$\varphi(\pm k, v) = \sum_{n=0}^{\infty} a_n V^n.$$

(6.13)

Substituting (6.13) into (6.12) and equating the factors at different degrees of V to zero, we obtain the recursion formula for a_n as follows:

$$a_{n+1} = \frac{(a^2V_0^2c + Vn(2n + 1 \pm 2ik_a))a_n - (n-1)na_{n-1}}{V_0^2(n+1)((n+1) \pm 2ika)}.$$

(6.14)

From the last formula, all the coefficients can be expressed by a_n. The value a_{n+1} can be found from the boundary conditions (6.3) and the expression (6.13)

$$\lim_{x\to\infty} \varphi(\pm k, V(x)) = a_0 = 1.$$

(6.15)

Now we can define the Jost function. From (6.1), we obtain

$$f(l, \pm k) = \varphi(\pm k, V_p)f_0(l, \pm k),$$

$$V_p = \lim_{x\to 0} V(x).$$

(6.16)

Using the known relationship [2] between the Jost function and the S-matrix element for s-scattering when $l = 0$, we obtain

$$S(k) = \frac{\varphi(+k, V_p)}{\varphi(-k, V_p)},$$

(6.17)

$$\varphi(+k, V_p) = \sum_{n=0}^{\infty} a_n(\pm k, V_p) V_p^n.$$

(6.18)

6.3   The Case of Large Nuclei

Making some transformation, Eq. (6.12) can be reduced to the form

$$\left(\left(1 - \frac{V}{V_0}\right)^2 - \left(1 - \frac{V}{V_0}\right)^3 \right) \frac{d^2\varphi(\pm k, V)}{dV^2}$$

$$+ \frac{1}{V_0^2} \left(2\left(1 - \frac{V}{V_0}\right)^2 - \left(1 - \frac{V}{V_0}\right) \pm 2ika\left(1 - \frac{V}{V_0}\right) \right)$$

$$\times \frac{d\varphi(\pm k, V)}{dV} - \frac{ca^2}{V}\varphi(\pm k, V) = 0.$$

(6.19)

Substituting the new variable

$$y = \ln\left(1 - \frac{V}{V_0}\right),$$

(6.20)

Eq. (6.19) can be rewritten as follows:

$$(1 - e^y)\frac{d^2\varphi(\pm k, V)}{dy^2} - (e^y \pm 2ika)\frac{d\varphi(\pm k, y)}{dy} - ca^2 V_0\varphi(\pm k, y) = 0.$$

(6.21)

To find the Jost function, we need to know the solution $\varphi(\pm k, V(x))$ at the origin. For the Woods–Saxon potentials (6.20), we have $\lim_{x \to 0} \exp\{y\} = 0$ as $V_p = V_0$. Then the last equation simplifies it as follows:

$$\frac{d^2\varphi(\pm k, y)}{dy^2} \mp 2ika\frac{d\varphi(\pm k, y)}{dy} - ca^2 V_0\varphi(\pm k, y) = 0.$$

(6.22)

Expression of the potential (6.10) shows that the approximation is correct in the region where

$$\exp\left\{\frac{x - R}{a}\right\} \ll 1. \tag{6.23}$$

Clearly, this depends on the region $0 \leq x \leq R$ of the nucleus and the parameter a values. Less than a, the x is closer to the parameter R, which is approximately equal to the radius of the core. The last equation is not difficult to solve in the following form:

$$\varphi_1(\pm k, y) = c_1(\pm k) \exp\left\{i(\pm\alpha + \sqrt{\alpha^2 \beta} y)\right\}$$

$$+ c_2(\pm k) \exp\left\{i(\pm\alpha + \sqrt{\alpha^2 \beta} y)\right\},$$

$$\alpha = ka, \quad \beta = a^2 cV. \tag{6.24}$$

In order to determine the unknown $c_1(\pm k)$ and $c_2(\pm k)$, it is necessary to know the exact solution of Eq. (6.21) and use the boundary condition (6.21). But we just solve the truncated equation

$$(1 \pm 2ika)\frac{d\varphi(\pm k, y)}{dy} - ca^2 V \varphi(\pm k, y) = 0. \tag{6.25}$$

This equation is obtained from the exact equation for larger distances, where the potential V takes small values. Then, according to (6.20) $y = 0$. Equation (6.25) is solved as follows:

$$\varphi_2(\pm k, y) = A \exp\left\{\frac{-a^2 cV_0 y}{1 \pm 2i\alpha}\right\}. \tag{6.26}$$

When $x \to \infty$, then $y \to 0$ and A is given as

$$\lim_{y \to 0} \varphi_2(\pm k, y) = A = 1. \tag{6.27}$$

We make both decisions at the point $x = R$, i.e., at the point where the potential is reduced by half. Of course, at this point, neither (6.24) nor (6.26) are precise:

$$\varphi_1\left(\pm k, y\big|_{y=\ln\frac{1}{2}}\right) = \varphi_2(\pm k, y)\big|_{y=\ln\frac{1}{2}},$$

$$\varphi_1'\left(\pm k, y\big|_{y=\ln\frac{1}{2}}\right) = \varphi_2'(\pm k, y)\big|_{y=\ln\frac{1}{2}}. \tag{6.28}$$

From the last four equations, we find:

$$c_1(\pm k) = \frac{1}{2}\exp\left\{\frac{\beta\ln 2}{1\pm 2i\alpha}\right\}\left(1\pm\frac{\alpha}{\sqrt{\alpha^2-\beta}} + \frac{i\beta}{\sqrt{\alpha^2-\beta}(1\pm 2i\alpha)}\right),$$

$$c_2(\pm k) = \frac{1}{2}\exp\left\{\frac{\beta\ln 2}{1\pm 2i\alpha}\right\}\left(1\pm\frac{\alpha}{\sqrt{\alpha^2-\beta}} - \frac{i\beta}{\sqrt{\alpha^2-\beta}(1\pm 2i\alpha)}\right). \tag{6.29}$$

Using the solution φ_2 and coefficients (6.29) from (6.28), the approximate expression of the scattering matrix was obtained

$$S|_{l=0} = \exp\left\{2i\alpha\left(-\frac{2\beta\ln 2}{1+2\alpha^2} + \ln 2 - \frac{R}{a}\right)\right\}Z,$$

$$Z = \frac{ctg\gamma - \dfrac{i\alpha}{\sqrt{\alpha^2-\beta}} - \dfrac{\beta(1-2i\alpha)}{\sqrt{\alpha^2-\beta(1+4\alpha^2)}}}{ctg\gamma + \dfrac{i\alpha}{\sqrt{\alpha^2-\beta}} - \dfrac{\beta(1+2i\alpha)}{\sqrt{\alpha^2-\beta(1+4\alpha^2)}}}, \tag{6.30}$$

$$\gamma = \sqrt{\alpha^2-\beta}\left(\ln 2 - \frac{R}{a}\right), \quad \alpha = ak, \quad \beta = ca^2 V_0.$$

For an incident neutron energy less than $10\,\text{keV}$, this formula can be simplified by neglecting the α_2 terms. In this case, the last formula simplifies as follows:

$$S = \exp\left\{2i\alpha\left((1-2\beta)\ln 2 - \frac{R}{a}\right)\right\}Z,$$

$$Z = \frac{ctg\gamma - \sqrt{-\beta} - \dfrac{i\alpha(1-2\beta)}{\sqrt{-\beta}}}{ctg\gamma + \sqrt{-\beta} + \dfrac{i\alpha(1+2\beta)}{\sqrt{-\beta}}}, \tag{6.31}$$

$$\gamma = \sqrt{-\beta}\left(\ln 2 - \frac{R}{a}\right).$$

Considering this formula as the main, we can apply it for complex potentials, where the functions are the same for the real and imaginary parts of the potential. Consider the case when the diffusion parameter a tends to zero. Then given that $\beta \to 0$, $\gamma = -\sqrt{cV_0}R$ (6.31), we obtain

$$S_0 = arctg\frac{kRtg\sqrt{-cV_0}}{\sqrt{-cV_0}} - kR. \qquad (6.32)$$

This result coincides with the formula in Ref. [3] for the case $V_0 \gg E$, as it should be, since the Woods–Saxon potential for small a differs little from the square well potential. Thus, the formula (6.30) is more accurate for small values of the diffusion parameter a, as shown in (6.23). It is mathematically equivalent to a large value R in the case of large nuclei.

6.4 Numerical Results and Conclusions

It should be noted that the series (6.18) converges rapidly for small nuclei with large values of diffusion coefficient. Calculations have shown that the scattering matrix obtained by the formula (6.17) is consistent with the results obtained using the software [4]. For example, as in Ref. [5] values for the parameters $R = 1.25A^{\frac{1}{3}}$ fm, $a = 0.65$ fm, $\text{Re}\,V_0 = -21.6$ MeV, $\text{Im}\,V_0 = 6.2$ MeV, $E_{\text{lab}} = 2$ MeV for scattering on Li_3^7 our calculations (6.17) $S = -0.245 - i0.615$, and [3] $S_j = -0.244 - i0.614$.

The small difference is obtained due to the finite number of terms by taking into account the practical calculations used in the formula (6.17). Of course, the biggest interest is in the approximate formula (6.30). This approximate formula gives quite good results for the energies of the incident neutrons in nuclei for the given values of the parameters.

From the above theory and (6.32), it follows that the approximate formula is accurate for small values of the diffusion parameter a, although in case of $a = 0.6$ fm for the energy $E = 0.10$ MeV and below they are sufficiently accurate.

Table 6.1. For the nucleus Pb^{208}, we used $a = 0.3\,\text{fm}$, $R = 8.354\,\text{fm}$ and $V_0 = -42\,\text{MeV}$.

$E \cdot \text{MeV lab.}$	S S_j	σ σ_j
0.001	$S = 0.9921 - i0.1252$ $S_j = 0.9920 - i0.1264$	$\sigma = 41.02$ $\sigma_j = 41.81$
0.01	$S = 0.9232 - i0.3843$ $S_j = 0.9208 - i0.3900$	$\sigma = 38.66$ $\sigma_j = 39.81$
0.1	$S = 0.2766 - i0.9610$ $S_j = 0.3018 - i0.9533$	$\sigma = 24.16$ $\sigma_j = 23.78$
0.5	$S = 0.9256 - i0.3785$ $S = 0.9421 - i0.3353$	$\sigma = 0.7493$ $\sigma_j = 0.5880$

Table 6.2. For nucleus Pb^{208}, we used $a = 0.6\,\text{fm}$, $R = 8.354\,\text{fm}$ and $V_0 = -42\,\text{MeV}$.

$E \cdot \text{MeV lab.}$	S S_j	σ σ_j 10^{-2} (barn)
0.001	$S = 0.9923 - i0.1236$ $S_j = 0.9917 - i0.1287$	$\sigma = 3997$ $\sigma_j = 4300$
0.01	$S = 0.9253 - i0.3791$ $S_j = 0.9185 - i0.3954$	$\sigma = 3762$ $\sigma_j = 4093$
0.1	$S = 0.3397 - i0.9405$ $S_j = 0.2851 - i0.9585$	$\sigma = 22314$ $\sigma_j = 2404$
0.5	$S = 0.9105 - i0.4136$ $S_j = 0.9517 - i0.3071$	$\sigma = 89.47$ $\sigma_j = 49.33$
2	$S = 0.5322 - i0.8466$ $S_j = 0.6460 - i0.7631$	$\sigma = 94.33$ $\sigma_j = 76.64$
60	$S = 0.8646 + i0.5020$ $S_j = -0.7575 + i0.6528$	$\sigma = 0.9547$ $\sigma_j = 1.611$
90	$S = 0.4072 - i0.9133$ $S_j = 0.4405 + i0.8977$	$\sigma = 2.437$ $\sigma_j = 2.354$
150	$S = 0.9476 - i0.3190$ $S_j = 0.9380 - i0.3453$	$\sigma = 0.1784$ $\sigma_j = 0.2091$

For energies that exceed the depth of the potential by two to three times, it turns out to be in quite good agreement with the program *JIB* even for $a = 0.6$ fm. It is evident from the results provided in Table 6.2. The results presented in Tables 6.1 and 6.2 using program *JIB* have an index *J*.

However, the main result of this paper should be considered using the fact that the wave function is a product of a function that depends on the interaction potential and a function describing the free motion of a particle. This makes it possible to replace the Schrödinger equation by Eq. (6.5).

References

1. Janawiczius, A. J. and Kwiatkowski, K. (1978). *Schrodinger Equation in Potential Representation for Saxon–Woods Potential Representation for the Case of S-Wave.* Report, No. 1001/PL, Krakow, p. 12.
2. de Alfaro, V. and Redge, T. (1966). *Potential Scattering.* Mir, Moscow, p. 66.
3. Sokolov, A. A. and Ternov, I. M. (1970). *Quantum Mechanics and Atomics Physics*, Prosveshchenie, Moscow, p. 213 (in Russian).
4. Perey, F. G. (1997). *Local optical program with automatically search of parameters.* Institute of Nuclear Physics, Krakow.
5. Xodgson, P. E. (1966). *Optical Model of Potential Scattering*, Atomizdat, Moscow, p. 97 (in Russian).

Chapter 7

A General Solution
of the Schrödinger Equation

7.1 Introduction

A general solution of the Schrödinger equation in the potential representation is put forward in the form of an integral equation A[2]. In this representation, the wave function can be expressed as a product of the free solution and the function which depends only on the interaction potential. Numerical calculations carried out for the Woods–Saxon potential are in good agreement with the scattering matrix elements which were obtained with the aid of an optical model code.

The wave function is proposed as a Jost solution that is a product of two parts

$$f(l, \pm k, x) = \varphi(l, \pm k, x) f_0(l, \pm k, x), \qquad (7.1)$$

where $f_0(l, \pm k, x)$ is the Jost solution in the absence of interaction A[4]. The interaction is taken into account in terms of the $\varphi(l, \pm k, x)$. In this way, $\varphi(x)$ depends only on the potential

$$\varphi(x) = \varphi(V(x)), \qquad (7.2)$$

and satisfies the following boundary condition:

$$\lim_{x \to \infty} \varphi(V(x)) = 1. \qquad (7.3)$$

Consider a function containing complete information about interaction and having convenient asymptotic and simple dependence on the wave number k and thus the energy that it differs from the Jost function. It has another remarkable property: the origin of it should be regular. It is easy to show, using the known task for the Jost function [2]

$$f(l + 1/2, k) = (2l + 1) \lim_{x \to 0} x^l f(l + 1/2, k, x). \qquad (7.4)$$

Then, from formula (7.1), we get

$$f(l + 1/2, k) = \varphi(l, k, 0) f_0(l + 1/2, k), \qquad (7.5)$$

where $f_0(l+1/2, k)$ is the Jost function in the absence of interaction. Using (7.5) $\varphi(l, k, x)$ and the scattering matrix expression A[2], [1] we get

$$S(l, k) = \frac{f(l + 1/2, k)}{f(l + 1/2, -k)} e^{i\pi \cdot l} \qquad (7.6)$$

after substituting

$$f_0(l + 1/2, k) = 2^{l+\frac{1}{2}} \left(\frac{2}{\pi}\right)^{\frac{1}{2}} \left((l + 3/2) k^{-1} \exp\left\{-il\frac{\pi}{2}\right\}\right),$$

$$f_0(l + 1/2, -k) = 2^{l+\frac{1}{2}} \left(\frac{2}{\pi}\right)^{\frac{1}{2}} \left((l + 3/2) k^{-1} \exp\left\{il\frac{\pi}{2}\right\}\right),$$

$$(7.7)$$

we get an expression of a scattering matrix including φ, i.e.,

$$S_l(k) = \frac{\varphi(l, k, 0)}{\varphi(l, -k, 0)}. \qquad (7.8)$$

The problem was solved only for the Woods–Saxon potential for the case of scattering A[1], and the results were more of theoretical rather than practical value. In this chapter, the problem is solved in a general form for any short-range potential. In this manner, we obtained the integral equations A[2] for practical use for the interpretation of nuclear scattering experiments.

7.2   General Solution

When the Schrödinger equation for the following potential representation (7.1)

$$\left(\frac{dV}{dx}\right)^2 \frac{d^2\varphi}{dV^2} + \left(\frac{d^2V}{dx^2} + 2\frac{f_0'}{f_0}\frac{dV}{dx}\right)\frac{d\varphi}{dV} - cV(x)\varphi = 0 \qquad (7.9)$$

is divided by $\frac{d\varphi}{dV}$, we obtain the equation

$$\left(\frac{dV}{dx}\right)^2 \frac{\frac{d^2\varphi}{dx^2}}{\frac{d\varphi}{dV}} + \frac{d^2V}{dx^2} + 2\frac{f_0'}{f_0}\frac{dV}{dx} - cV(x)\frac{\varphi}{\frac{d\varphi}{dV}} = 0. \qquad (7.10)$$

Next, dividing it by $\frac{dV}{dx}$, we obtain

$$\frac{\frac{d^2V}{dx^2}}{\frac{dV}{dV}} + 2\frac{f_0'}{f_0} = cV(x)\frac{\varphi}{\frac{dV}{dx}\frac{d\varphi}{dV}} - \left(\frac{dV}{dx}\right)^2 \frac{\frac{d^2\varphi}{dV^2}}{\frac{d\varphi}{dV}} = 0. \qquad (7.11)$$

Rewriting the resulting expression in the form of derivatives of logarithms, we obtain

$$\left(\ln\frac{dV}{dx}\right)' x + (\ln f_0^2)' x = \frac{cV}{(\ln\varphi)'x} - \left(\ln\frac{d\varphi}{dV}\right)' x. \qquad (7.12)$$

Providing the transformation

$$\left(\ln\frac{d\varphi}{dV}\frac{dV}{dx}f_0^2\right)' x = \frac{cV}{(\ln\varphi)'x}, \qquad (7.13)$$

we obtain the possibility to express the potential

$$V = \frac{1}{c}(\ln\varphi)'x\left(\ln\frac{d\varphi}{dx}f_0^2\right)' x. \qquad (7.14)$$

Taking the logarithms of these derivatives and reducing them to φ_x, we obtain

$$V = \frac{1}{c}\frac{(\varphi_x' f_0^2)'x}{\varphi \cdot f_0^2}. \qquad (7.15)$$

Thus, we find that the potential is very simply expressed in terms of φ.

Performing a simple transformation and after integrating, we obtain the following important expression:

$$\varphi_x' f_0^2 = c \int V\varphi \cdot f_0^2 dx + a. \tag{7.16}$$

We divide this expression by $f_0^2(l, k, x)$. After integration of this expression, we obtain an integral equation,

$$\varphi(l, k, x) = \int \frac{cdx}{f_0(l, k, x)} dx \int V f_0^2 dx + \int \frac{adx}{f_0^2} + b. \tag{7.17}$$

Using the boundary conditions

$$\lim_{x \to \infty} \varphi = 1, \tag{7.18}$$

$$\lim_{x \to \infty} \varphi_x' = 0 \tag{7.19}$$

imposed on φ and by applying $\lim_{x \to \infty} V = 0$ for (7.16) and (7.17), we obtain the values of the constants of integration: $a = 0$, $b = 1$. Then we can rewrite (7.17) and get a very useful integral equation for practical calculations of nuclei scattering:

$$\varphi(l, k, x) = c \int_\infty^x \frac{dz}{f_0^2(l, k, z)} \int_\infty^z V(y)\varphi(l, k, y) f_0^2(l, k, y) dy + 1. \tag{7.20}$$

From this formula, it is not difficult to obtain the scattering matrix (7.8) with high accuracy. Let us start with the following equations:

$$V(y)\varphi(y) f_0^2(y) = \frac{d\Phi}{dy}, \tag{7.21}$$

$$\lim_{z \to \infty} \frac{d\Phi(z)}{dz} = 0 \tag{7.22}$$

$$\lim_{z \to \infty} \Phi(z) = \text{Const.} \tag{7.23}$$

Using (7.21), (7.22) and (7.23), the formula (7.16) can be rewritten as follows:

$$\varphi_z' f_0^2 = c \int_\infty^z V(y)\varphi(l, k, y) f_0^2(l, k, y) dy = \Phi(z) - \Phi(\infty). \tag{7.24}$$

Comparing this formula with (7.24), we see that

$$a = -\Phi(\infty). \tag{7.25}$$

If $a = 0$, then we can write $\tag{7.26}$

$$\varphi_z'(l,k,z) = \frac{1}{f_0^2(l,k,z)} \int_\infty^z V(y)\varphi(l,k,y)f_0^2(l,k,y)dy. \tag{7.27}$$

Integrating this task and using (7.18), we obtain (7.20). This solution satisfies the following Schrödinger equation (7.1):

$$f_0\frac{d^2\varphi}{dx^2} + 2f_0'\frac{d\varphi}{dx} - cVf_0\varphi = 0, \tag{7.28}$$

which is obtained from the ordinary differential equation by substituting (7.1). Due to direct substitution of (7.20) in (7.28), it is easy to verify that (7.20) is an integral equation for finding the function $\varphi(l,\pm k,x)$. For the zero approximation (7.20), we take $\varphi_0 = 1$. Integrating according to (7.20), we can find the second approach φ_1. Substituting φ_1 into the right-hand side of (7.20), we obtain φ_2 and continue with the following approximations of $\varphi(l,k,x)$. Merging (7.20) and (7.8), we obtain the resulting expression for the S-matrix as follows:

$$S_l(k) = \frac{c\int_\infty^0 \frac{dz}{f_0^2(l,k,z)} \int_\infty^z V(y)\varphi(l,k,y)f_0^2(l,k,y)dy + 1}{c\int_\infty^0 \frac{dz}{f_0^2(l,-k,z)} \int_\infty^z V(y)\varphi(l,-k,y)f_0^2(l,-k,y)dy + 1},$$

$$c = \frac{2m}{\hbar^2}, \qquad \mu = \frac{m_1 m_2}{m_1 + m_2}. \tag{7.29}$$

For practical explicit expressions of free Jost solutions [2], $f_0(l,\pm k,x)$ should be used.

7.3   Numerical Results and Conclusions

Calculations of neutrons scattering on nuclei Fe^{59}, Zr^{90} and Pb^{208} using the optical model [1, 3] showed good agreement with the results obtained using the program [4] JIB. Scattering matrix elements are in good agreement for all partial waves. This is evident from the data in Tables 7.1–7.3 for orbital moments $l = 0, 1, 2, \ldots$. Thus, the postulate about the possibility of representing the wave function as

Table 7.1. S-matrices obtained by the potential representation for neutrons scattering at Fe^{59} in the case of Woods–Saxon potential with presented parameters A[2] comparison after n iterations with standard calculations [4] of S_{JIB}, $u = -39$ MeV, $W = -7.8$ MeV, $a = 0.35$ fm, $R_0 = 1.45$ fm, $E_{lab} = 7$ MeV.

l	S	S_{JIB}	n
0	$0.1401 - i0.2631$	$0.1403 - i0.2634$	15
1	$-0.3715 + i0.1040$	$-0.3716 + i0.1026$	14
2	$-0.3713 - i0.3065$	$-0.3700 - i0.3069$	13
3	$0.3987 - i0.1374$	$0.3983 - i0.1358$	13
4	$0.8593 - i0.1171$	$0.8581 - i0.1154$	12
5	$0.9234 - i0.0117$	$0.9236 - i0.0113$	12

Table 7.2. S-matrix calculated by potential representation for neutron scattering at Zr^{90} in the case of Woods–Saxon potential A[2] with the comparison of the presented parameters after n iterations with [4] standard calculations of S_{JIB}, $u = -45$ MeV, $W = -4.5$ MeV, $a = 0.5$ fm $R_0 = 1.38$ fm, $E_{lab} = 7$ MeV.

l	S	S_{JIB}	n
0	$0.4653 - i0.2546$	$0.4655 - i0.2554$	16
1	$0.3567 - i0.0422$	$0.3567 - i0.0417$	15
2	$-0.1488 - i0.5227$	$-0.1485 - i0.5229$	15
3	$-0.1102 - i0.5589$	$-0.1094 - i0.5589$	15
4	$0.5972 + i0.1967$	$0.5968 + i0.1977$	15
5	$0.9499 - i0.0195$	$0.9498 - i0.0186$	14
6	$0.9698 + i0.0234$	$0.9698 + i0.0237$	13
7	$0.9996 + i0.0016$	$0.9996 + i0.0017$	12

a product of two parts, one of which depends on the interaction potential and the other describes the free motion, can be considered to have been proven. It may prove fruitful in solving problems related to the application of group theory in quantum mechanics and the inverse scattering theory. The formulas obtained as integral equations can be used for fitting the parameters to the potential of experimental scattering data and also for calculating the exact values of the scattering matrixes with a less exact standard program [4] for the optical model.

Table 7.3. S-matrix obtained by the potential representation for neutron scattering at Pb^{208} in the case of Woods–Saxon potential with the comparison of the presented parameters after n iterations with standard calculations of S_{JIB}, $u = -28.2\,\text{MeV}$, $W = -11\,\text{MeV}$, $a = 0.6\,\text{fm}$, $R_0 = 1.25\,\text{fm}$, $E_{lab} = 80\,\text{MeV}$.

l	S	S_{JIB}	n
0	$0.0166 - i0.1746$	$0.0171 - i0.1746$	14
1	$0.0138 - i0.1774$	$0.0141 - i0.1770$	14
2	$0.0065 - i0.1783$	$0.0043 - i0.1784$	14
3	$-0.0047 - i0.1833$	$-0.0190 - i0.1830$	14
4	$-0.0192 - i0.1852$	$-0.0190 - i0.1853$	13
5	$-0.0420 - i0.1887$	$-0.0414 - i0.1885$	13
6	$-0.0663 - i0.1894$	$-0.0661 - i0.1893$	13

At present, it is still difficult to say with certainty about the prospects of application of the new method in practical calculations, but the results are encouraging. Replacing the function $f_0(l, \pm k, x)$ on the spherical Coulomb functions A[6], this problem can be easily generalized for the scattering of charged particles.

We can generalize the presented postulate: with the addition of a new potential to the Hamiltonian there appears a new function, and multiplying that by the old, we get a new wave function.

References

1. Devanarayanan, S. (2016). *A Text Book on Nuclear Physics*, Lexington, USA, p. 380.
2. de Alfaro, V. and Regge, T. (1965). *Potential Scattering*, North-Holland Publishing Company, Amsterdam, p. 274.
3. Jelley, N. J. (1990). *Fundamentals of Nuclear Physics*, Cambridge University Press, Cambridge, p. 278.
4. Perey, F. G. (1977). *Local Optical Program with Automatic Parameters Search*, Krakow Computing Center.

Chapter 8

The General Solutions for Positive and Negative Energies

8.1 Introduction

A general solution of the Schrödinger equation in the potential representation is put forward in the form of integral equations. In this representation, the wave function is expressed as a product of the unperturbed solution and the function which depends on the interaction potential A[1]. Numerical calculations carried out for the Woods–Saxon potential are in good agreement with the scattering matrix elements *which were* obtained with the aid of an optical model code. The formulas for the bound state perturbation task and relativistic corrections in the optical model are obtained.

In this chapter, the analytical solutions of the one-particle radial Schrödinger equation for positive and negative energies are obtained by using the potential representation method which was proposed in Ref. A[4]. The main idea of this method is expressing the radial wave function in terms of the potential A[1]. For positive energies and real k, the Jost function is expressed as a product A[1, 2]

$$f(l, -k, r) = \varphi(l, -k, r) f_u(l, -k, r)$$
$$f(l, -k, r \sim \infty).\exp(-ikr) = 1 \tag{8.1}$$

of the function φ which depends on the perturbation potential $V(r)$ and the radial unperturbed Jost solution of the Schrödinger equation for the unperturbed potential $V_0(r)$. The function φ has a very handy asymptotic expression

$$\lim_{r \to \infty} \varphi(l, -k, r) = 1, \tag{8.2}$$

which does not depend on the wave number k and energy. The relation (8.1) was derived in Ref. A[1] using the undefined Lagrange coefficient method [1, 2]. Recently, condition (8.1) for the perturbative scattering phase shift convergence was used in Ref. A[1].

Substituting Eq. (8.1) into the one-particle radial Schrödinger equation

$$\frac{d^2}{dr^2}f + \left(k^2 - \frac{l(l+1)}{r^2} - cV(r) - cV_0(r)\right) f = 0,$$

$$c = \frac{2m}{\hbar^2}, \quad k^2 = \frac{2mE}{\hbar^2}, \tag{8.3}$$

and taking into account that f_u is the Jost solution for the potential $V_0(r)$, we can obtain the radial Schrödinger equation in the potential representation

$$f_u \frac{d^2\varphi}{dr^2} + 2\frac{df_u}{dr}\frac{d\varphi}{dr} - cf_u V(x)\varphi = 0. \tag{8.4}$$

This differential equation can be easily transformed into the integral equation A[1]. The potential scattering phase shifts are defined by Jost solutions at the origin. Even when the free Jost solution instead of the unperturbed solution was taken, the convergence of the phase shifts was obtained in Ref. A[1], but perturbation in this case was not small. The program JIB [3] for the phase shift calculations, where the standard discretization method [2] was used, is more complicated and less exact.

The integral equations for calculating the perturbed radial wave functions and eigen energies are obtained in this chapter by using the unperturbed irregular solutions instead of the Jost solutions. A similar method for the ground states was proposed previously by Zeldovich [3].

8.2 The Integral Equation for Positive Energies in the Potential Representation

First of all, we consider the solution of Eq. (8.4) for positive energies. By integrating (8.4), we obtain

$$\varphi(l,k,r) = c \int \frac{1}{f_u^2} dr \int V f_u^2 \varphi dr + \int a \frac{1}{f_u^2} dr + b. \qquad (8.5)$$

Using the boundary conditions

$$\lim_{r \to \infty} \varphi = 1, \qquad (8.6)$$

$$\lim_{r \to \infty} \frac{d}{dr} \varphi = 0, \qquad (8.7)$$

we can get

$$a = 0, \quad b = 1. \qquad (8.8)$$

Taking into account these conditions and Eq. (8.4), we have A[2]

$$\varphi(l,k,r) = c \int_\infty^r \frac{1}{f_u^2(l,k,y)} dy \int_\infty^y V f_u^2(l,k,x) \varphi(l,k,x) dx + 1. \qquad (8.9)$$

A similar equation is obtained and considered later in Ref. A[2].

Using the well-known definition for the Jost function [1]

$$f(l,k) = (2l+1) \lim_{r \to \infty} r^l f(l,k,r), \qquad (8.10)$$

the eigenvalues of the scattering matrix can be presented [6] by Jost functions

$$S(l,k) = \frac{f(l,k)}{f(l,-k)} e^{i\pi l}. \qquad (8.11)$$

From the free Jost functions [6]

$$f_0(l,k) = \frac{2^{l+1}}{\sqrt{\pi}} \Gamma \left(l + \frac{3}{2} \right) k^{-1} \exp \left\{ -i\frac{\pi}{2}l \right\}, \qquad (8.12)$$

$$f_0(l,-k) = \frac{2^{l+1}}{\sqrt{\pi}} \Gamma \left(l + \frac{3}{2} \right) k^{-1} \exp \left\{ -i\frac{\pi}{2}l \right\}, \qquad (8.13)$$

as well as Eqs. (8.1) and (8.10)–(8.13), the expression of the scattering matrix has been obtained as follows:

$$S_l(k) = \frac{\varphi(l, k, 0)}{\varphi(l, -k, 0)}. \tag{8.14}$$

Using this formula, Eq. (8.9) and free Jost functions (8.12), (8.13), the following expression has been obtained:

$$S_l = \left(c \int_\infty^0 \frac{1}{f_u^2(l, k, x)} dx \int_\infty^x V f_u^2(l, k, y) \varphi(l, k, y) dy + 1 \right)$$
$$\times \left(c \int_\infty^0 \frac{1}{f_u^2(l, -k, x)} dx \int_\infty^x V f_u^2(l, -k, y) \varphi(l, -k, y) dy + 1 \right)^{-1}. \tag{8.15}$$

In this case, the free solutions were used as unperturbed solutions. For obtaining the unperturbed Jost functions, we must solve the integral Eq. (8.9) for the unperturbed potential. In this case, we must use the free Jost solutions [6] expressed by Hankel functions

$$f_0(l, k, x) = \left(\frac{\pi}{2} \right)^{\frac{1}{2}} \exp\left\{ -i\frac{\pi}{2}(l+1) \right\} (kx)^{\frac{1}{2}} H_{l+\frac{1}{2}}^{(2)}(kx), \tag{8.16}$$

$$f_0(l, -k, x) = \left(\frac{\pi}{2} \right)^{\frac{1}{2}} \exp\left\{ i\frac{\pi}{2}(l+1) \right\} (kx)^{\frac{1}{2}} H_{l+\frac{1}{2}}^{(1)}(kx). \tag{8.17}$$

Substituting Eq. (8.9) into Eq. (8.14) and using Eqs. (8.1), (8.10) and (8.11), we can obtain the recursion relation for the scattering matrix S_t and $S_{l,u}$ elements

$$S_l = \left(c \int_\infty^0 \frac{1}{f_u^2(l, k, x)} dx \int_\infty^x V f_u^2(l, k, y) \varphi(l, k, y) dy + 1 \right)$$
$$\times \left(c \int_\infty^0 \frac{1}{f_u^2(l, -k, x)} dx \int_\infty^x V f_u^2(l, -k, y) \varphi(l, -k, y) dy + 1 \right)^{-1}$$
$$\times S_{l,u}, \tag{8.18}$$

for unperturbed and perturbed potentials, respectively.

In Ref. A[14], it was shown that the relativistic corrections to mass must be included in the optical and shell model calculations.

Then using the perturbation of differential semi-relativistic nucleon mass correction operator D A[4, 15], we get

$$DF_l = \frac{d^4}{dy^4}F_l - \frac{2L_0}{y^2}\frac{d^2}{dy^2}F_l + \frac{4L_0}{y^3}\frac{d}{dr}F_l + \frac{L_0^2 - 6L_0}{y^4}F_l,$$
(8.19)

$$L_0 = L(L+1), F_l(k, y) = \varphi(l, k, y)f_u(l, k, y).$$

From Eqs. (8.18) and (8.19), we can get the expression for the semi-relativistic scattering as follows:

$$S_l = \left(C_1 \int_\infty^0 \frac{-1}{f_u^2(l, k, x)}dx \int_\infty^x f_u(l, k, y)DF_l(k, y)dy + 1\right)$$

$$\times \left(C_l \int_\infty^0 \frac{-1}{f_u^2(l, -k, x)}dx \int_\infty^x f_u(l, -k, y)DF_l(-k, y)dy + 1\right)^{-1}$$

$$\times S_{l,u},$$
(8.20)

where $C_1 = \left(\frac{\hbar}{2mc}\right)^2$, and the semi-relativistic perturbation function

$$\varphi(l, k, r) = C_1 \int_\infty^r \frac{-1}{f_u^2(l, k, y)}dy \int_\infty^y f_u(l, k, x)DF_l(k, x)dx + 1.$$
(8.21)

8.3 The Integral Equation for Negative Energies in the Potential Representation

The radial eigenfunctions in the potential representation can be expressed as a product of the unperturbed wave function $u_{\alpha u}(r)$ and multiplier $\varphi_\alpha(r)$, which depend on the perturbation potential $V(r)$:

$$U_\alpha(r) = \varphi_\alpha(r)u_{\alpha u}(r).$$
(8.22)

We suppose that the unperturbed eigenvalue problem

$$\frac{d^2}{dr^2}u_{\alpha u}(r) - \frac{l(l+1)}{r^2}u_{\alpha u}(r) + c(E_{\alpha U} - V_0(r))u_{\alpha u}(r) = 0$$
(8.23)

has already been solved. The Schrödinger equation which we seek to solve is

$$\frac{d^2}{dr^2}U_\alpha(r) - \frac{l(l+1)}{r^2}U_\alpha(r) + c(E_{\alpha U} + \Delta E_\alpha - V_0(r) - V(r))U_\alpha(r) = 0. \tag{8.24}$$

Substituting Eq. (8.22) into Eq. (8.24) and taking into account Eq. (8.23), we get the Schrödinger equation in the potential representation for the bound states

$$u_{\alpha u}\frac{d^2}{dr^2}\varphi_\alpha + 2\left(\frac{d}{dr}u_{\alpha u}\right)\frac{d}{dr}\varphi_\alpha + c(\Delta E_\alpha - V(r))u_{\alpha u}\varphi_\alpha = 0. \tag{8.25}$$

Multiplying the obtained equation by $u_{\alpha u}$, we have

$$\frac{d}{dr}\left(u_{\alpha u}^2\frac{d}{dr}\varphi_\alpha\right) = -cu_{\alpha u}(\Delta E_\alpha - V(r))u_{\alpha u}\varphi_\alpha. \tag{8.26}$$

Using the boundary condition

$$\lim_{r \to 0}\varphi_\alpha = 1, \tag{8.27}$$

we get the following equation after integration:

$$\varphi_\alpha = 1 - c\int_0^r \frac{1}{u_{\alpha u}^2}dy \int_0^y u_{\alpha u}(\Delta E_\alpha - V(z))u_{\alpha u}\varphi_\alpha dz. \tag{8.28}$$

The latter integral must be calculated by multiplying Eq. (8.28) by the unperturbed solution $u_{\alpha u}$, because in the nodes $u_{\alpha u}$ we have infinities for φ_α.

For bound states, we must require

$$\lim_{r \to \infty}\varphi_\alpha(r)u_{\alpha u}(r) = 0. \tag{8.29}$$

Then for attractive perturbation potentials, we must have consequently

$$\lim_{r \to \infty}\varphi_\alpha(r) = 0, \tag{8.30}$$

$$\lim_{r \to \infty}\varphi_\alpha(r) = \infty. \tag{8.31}$$

It is easy to verify that the conditions (8.29)–(8.31) will be satisfied in Eq. (8.28) if

$$\Delta E_\alpha = \frac{\int_0^\infty u_{\alpha u} V(r) \varphi_\alpha u_{\alpha u} dr}{\int_0^\infty u_{\alpha u} \varphi_\alpha u_{\alpha u} dr}, \quad E_\alpha = E_{\alpha u} + \Delta E_\alpha. \quad (8.32)$$

By simultaneously solving Eqs. (8.28) and (8.32) by the iteration method, we can obtain the eigenfunctions $U_\alpha(r)$ and eigenvalues E_α. For the first iteration, we must take $\varphi_\alpha = 1$, $\Delta E_\alpha = 0$.

8.4 Numerical Results and Conclusions

Providing calculations of the neutron scattering from nuclei Fe^{59}, Zr^{90} and Pb^{208} with Eqs. (8.9) and (8.15) in a wide range of energies, we obtain A[1] the elements of the scattering matrix S. The results presented in Tables 8.1–8.3 are in good agreement with the results obtained in Refs. A[1, 2] by using the optical model and those calculated with the program JIB [1]. These results show the possibility of using the potential representation method in which the perturbed eigenfunction can be obtained by multiplying the unperturbed eigenfunction by a factor which depends on the perturbation potential. The relation (8.18) obtained for the scattering matrices S_t and $S_{l,u}$ for perturbed and unperturbed potentials can

Table 8.1. S-matrix for the scattering of neutrons with energy $E_{\text{lab}} = 7\,\text{MeV}$ in laboratory system on nuclei Fe^{59} in the case of Woods–Saxon potentials with the parameters $V = -39\,\text{MeV}$, $W = -7.8\,\text{MeV}$, $\alpha = 0.35\,\text{fm}$ and $R = 5.64\,\text{fm}$ and n is the number of iterations.

l	S_l	S_{JIB}	n
0	$0.1401 - i\,0.2631$	$0.1403 - i\,0.2634$	15
1	$-0.3715 + i\,0.1040$	$-0.3716 - i\,0.1026$	14
2	$-0.3713 - i\,0.3065$	$-0.3700 - i\,0.3069$	13
3	$0.3987 - i\,0.1374$	$0.3983 - i\,0.1358$	13
4	$0.8593 - i\,0.1171$	$0.8581 - i\,0.1154$	12
5	$0.9234 - i\,0.0117$	$0.9236 - i\,0.0113$	13

Table 8.2. *S*-matrix for the scattering of neutrons with energy $E_{\text{lab}} = 7\,\text{MeV}$ on Zr^{90} in the case of Woods–Saxon potentials with the parameters $V = -45\,\text{MeV}$, $W = -4.5\,\text{MeV}$, $\alpha = 0.5\,\text{fm}$ and $R = 6.18\,\text{fm}$, and n is the number of iterations.

l	S_l	S_{JIB}	n
0	$0.4653 - i\,0.2546$	$0.4655 - i\,0.2554$	16
1	$0.3567 - i\,0.0422$	$0.3567 - i\,0.0417$	15
2	$-0.1488 - i\,0.5227$	$-0.1485 - i\,0.5229$	15
3	$-0.1102 - i\,0.5589$	$-0.1094 - i\,0.5589$	15
4	$0.5972 + i\,0.1967$	$0.5968 + i\,0.1977$	15
5	$0.9499 - i\,0.0195$	$0.9498 - i\,0.0186$	14
6	$0.9698 + i\,0.0234$	$0.9698 + i\,0.0237$	13
7	$0.9996 + i\,0.0016$	$0.9996 + i\,0.0017$	12

Table 8.3. *S*-matrix for the scattering of neutrons with energies $E_{\text{lab}} = 7\,\text{MeV}$ on Pb^{208} in the case of Woods–Saxon potentials with the parameters $V = -28.2\,\text{MeV}$, $W = -11\,\text{MeV}$, $\alpha = 0.6\,\text{fm}$ and $R = 7.41\,\text{fm}$, and n is the number of iterations.

l	S_l	S_{JIB}	n
0	$0.0166 - i\,0.1746$	$0.0171 - i\,0.1746$	14
1	$0.0138 - i\,0.1774$	$0.0141 - i\,0.1770$	14
2	$0.0065 - i\,0.1783$	$0.0043 - i\,0.1784$	14
3	$-0.0047 - i\,0.1833$	$-0.0190 - i\,0.1830$	14
4	$-0.0192 + i\,0.1852$	$-0.0190 + i\,0.1853$	13
5	$-0.0420 - i\,0.1887$	$-0.0414 - i\,0.1885$	13
6	$-0.0663 - i\,0.1894$	$-0.0661 - i\,0.1893$	13

be used for separation of the acting forces. It is useful for solving the inverse scattering problem when the investigation of the perturbation potential is needed.

Numerical calculations carried out with Eqs. (8.9) and (8.15) for the Woods–Saxon potential are provided with sufficient accuracy even if the number of iterations is less than 17. Here, the free solutions were considered to be unperturbed and we got convergence of the iteration process fast enough. The equivalent integral equations for finding eigenvalues of the scattering matrix have been found with the help of the modified method of indefinite coefficients [2]. The solutions of these integral equations are always more exact than the methods where numerical differentiation is used [3, 4].

In case we want to include relativistic corrections into the phase shifts, we must change $C_1 D$ in Eqs. (8.20) and (8.21) by $C_1 D - cV$ and insert $S_{l,u} = 1$ into Eq. (8.20).

For negative energies, the presence of the nodes in the unperturbed functions $u_{\alpha u}$ induces the poles in φ_α (8.28). This problem was solved in Ref. [6] by providing iterations for the product $\varphi_\alpha \varphi_{\alpha u}$ when solving a similar integral equation. In the standard logarithmic perturbation method A[14], this problem is more complicated. For avoiding this complication, we must multiply Eq. (8.28) by the unperturbed functions $u_{\alpha u}$.

References

1. Au, C. K., Chi-Keung Chow, and Chang-Sun Chu (1997). Perturbative scattering phase shifts in one dimension: Closed form results, *Phys. Lett. A*, **226**, 327.

2. Guter, R. S. and Janpolski, A. R. (1976). *Differential Equations*, Vyssheya Shkola, Moscow, p. 304.

3. Perey, F. G. (1977). *Local Optical Program with Automatical Parameters Search*, Krakow Computing Center.

4. Ortega, J. M. and Pool, V. G. (1981). *An Introduction to Numerical Methods for Differential Equations*, New York, Pitman Publishing Inc., p. 288.

5. Zeldovich, J. B. (1985). Perturbation theory for one dimension task in quantum mechanics and method Lagrange, *JETP*, **31**, 1101.

6. de Alfaro, V. and Regge, T. (1965). *Potential Scattering*, North-Holland Publishing Company, Amsterdam, p. 274.

7. Turbiner, A. B. (1984). Task about specter in quantum mechanics and procedure of nonlinearization, *Adv. Phys. Sci.*, **144**(1), 35.

Chapter 9

The Connection between Scattering Matrices for Different Potentials

9.1 Introduction

Using a modified method, undefined coefficients were found using integral equations for short-radii potentials for positive energies. The relation between scattering matrix and concrete definite potential was also identified A[4]. The kernel of integral equations are the Green's functions.

In this chapter, the potential representation method was used A[2] which was grounded on factorization A[1, 11] of the wave function when the interaction potential is added. We will consider the common properties of the solutions of the Schrödinger equation

$$\frac{d^2u}{dx^2} + (k^2 - cV(x) - V_0(x))u = 0, \qquad (9.1)$$

where

$$c = \frac{2m}{\hbar^2}, \qquad (9.2)$$

$$V_0(x) = cV_1(x) + \frac{l(l+1)}{x^2}. \qquad (9.3)$$

We will use potentials $V(x)$, $V_1(x)$ with asymptotes at zero $\lim_{x \to 0} V \approx \frac{1}{x^\alpha}, \alpha < 2$ that decrease faster at infinities than $\frac{1}{x^2}$. Jost solution A[1] can be expressed by $f_0(l, \pm k, x)$ for $V(x) = 0$ and the multiplier $\varphi(l, \pm k, x)$ depending on $V(x)$, i.e.,

$$f(l, \pm k, x) = \varphi_1(l, \pm k, x) f_0(l, \pm k, x). \tag{9.4}$$

Then Eq. (9.1) acquires the following form A[1]:

$$f_0 \frac{d^2 \varphi_1}{dx^2} + 2 \frac{df_0}{dx} \frac{d\varphi_1}{dx} - f_0 cV(x) \varphi_1 = 0. \tag{9.5}$$

The function φ_1 when interaction disappears must satisfy the requirement $\lim\limits_{V \to 0} \varphi_1 = 1$.

9.2   Integral Equations for Positive Energies

Here, we will obtain the integral equations using the method of undefined coefficients for positive energies $k^2 = 2mE/\hbar^2 > 0$

We will suppose that linearly independent solutions of (9.1) for the case $V(x) = 0$ satisfy the following asymptotes [1]:

$$\lim_{x \to 0} \varphi_0(l, k, x) = x^{l+1}, \tag{9.6}$$

$$\lim_{x \to \infty} f_0(l, \pm k, x) = e^{\mp ikx}. \tag{9.7}$$

Using the method of indefinite coefficients [2], the solution of (9.1) can be represented as

$$f = c_1(x) f_0 + c_2(x) \varphi_0, \tag{9.8}$$

where by applying factorization (9.4) we obtain

$$f = \left(c_1(x) + c_2(x) \frac{\varphi_0}{f_0} \right) f_0. \tag{9.9}$$

From this, we get the following expressions for the function:

$$\varphi_1 = c_1(x) + c_2(x) \frac{\varphi_0}{f_0}, \tag{9.10}$$

and its derivative

$$\varphi_1' = c_2(x) \frac{f_0 \varphi_0' - \varphi_0 f_0'}{f_0^2}. \tag{9.11}$$

Taking account of the additional condition

$$c_1'(x) + c_2'(x) \frac{\varphi_0}{f_0} = 0, \tag{9.12}$$

we obtain the second derivative

$$\varphi_1'' = c_2'(x)\frac{f_0(l,k)}{f_0^2} - 2c_2(x)\frac{f_0(l,k)}{f_0^2}f_0', \tag{9.13}$$

and Jost function

$$f_0(l,k) = f_0\varphi_0' - \varphi_0 f_0'. \tag{9.14}$$

After substituting (9.13) and (9.11) into (9.5) and taking into account (9.14), we have

$$c_2'(x)\frac{f_0(l,k)}{f_0} - cf_0 V\varphi_1 = 0. \tag{9.15}$$

After integration, we find $c_2(x)$ to be

$$c_2(x)\frac{1}{f_0(l,k)}\int_\infty^x cf_0^2 V\varphi_1 dx' \tag{9.16}$$

with the following boundary conditions included:

$$\lim_{x\to\infty} f(l,\pm k,x) = \lim_{x\to\infty} f_0(l,\pm k,x),$$
$$\lim_{x\to\infty} \varphi_1(l,\pm k,x) = 1. \tag{9.17}$$

From (9.8), we obtain

$$\lim_{x\to\infty} c_2(x) = 0. \tag{9.18}$$

From (9.12) and (9.15), the following derivative is obtained:

$$c_1'(x) = -\frac{1}{f_0(l,k)}cV\varphi_0 f_0\varphi_1. \tag{9.19}$$

After integrating the coefficient, we obtain

$$c_1(x) = -\frac{1}{f_0(l,k)}\int_0^x cV f_0\varphi_0\varphi_1 dx' + \alpha \tag{9.20}$$

by taking into account the condition

$$\lim_{x\to 0} c_1(x) = \alpha. \tag{9.21}$$

Then from (9.16), (9.20) and (9.8), we get

$$\alpha = 1 + \frac{1}{f_0(l,k)}\int_0^\infty cV f_0\varphi_0\varphi_1 dx'. \tag{9.22}$$

From the estimated coefficients $c_1(x)$ and $c_2(x)$ in compliance with (9.10), we can obtain the solution

$$\varphi_1 = \alpha - \frac{1}{f_0(l,k)} \int_0^x cV f_0 \varphi_0 \varphi_1 dx' - \frac{1}{f_0(l,k)} \frac{\varphi_0}{f_0} \int_x^\infty cV f_0^2 \varphi_1 dx'$$

$$(9.23)$$

for potential representation Eq. (9.5), which can be verified after substitution.

We must bear in mind that limits of integration in (9.16) and (9.20) were found under boundary conditions; however, the directions of integration can be defined only after substitution in (9.5).

Similarly, we can find the solution

$$\varphi(l,k,x) = \varphi_2(l,k,x)\varphi_0(l,k,x). \qquad (9.24)$$

of the Schrödinger equation in potential representation

$$\varphi_0 \frac{d^2\varphi_2}{dx^2} + 2\frac{d\varphi_0}{dx}\frac{d\varphi_2}{dx} - \varphi_0 cV\varphi_2 = 0 \qquad (9.25)$$

for the boundary condition

$$\lim_{x\to 0} \varphi_2(l,k,x) = 1, \qquad (9.26)$$

following from the boundary conditions:

$$\lim_{x\to 0} \varphi = x^{l+1}, \qquad (9.27)$$

$$\lim_{x\to 0} \varphi_0 = x^{l+1} \qquad (9.28)$$

$$\varphi_2 = \beta - \frac{1}{f_0(l,k)} \frac{f_0}{\varphi_0} \int_0^x cV \varphi f_0^2 \varphi_2 dx'$$

$$- \frac{1}{f_0(l,k)} \int_x^\infty cV \varphi_0 f_0 \varphi_2 dx',$$

$$\beta = 1 + \frac{1}{f_0(l,k)} \int_0^\infty cV \varphi_0 f_0 \varphi_2 dx'. \qquad (9.29)$$

The obtained Eqs. (9.23) and (9.29) can be rewritten in the following manner:

$$\varphi_1 = \frac{1}{f_0} \left(f_0 \alpha - \frac{f_0}{f_0(l,k)} \int_0^x cV f_0 \varphi_0 \varphi_1 dx' \right.$$

$$\left. - \frac{\varphi_0}{f_0(l,k)} \int_x^\infty cV f_0^2 \varphi_1 dx' \right), \tag{9.30}$$

$$\varphi_2 = \frac{1}{\varphi_0} \left(\varphi_0 \beta - \frac{f_0}{f_0(l,k)} \int_0^x cV \varphi_0^2 \varphi_2 dx' \right.$$

$$\left. - \frac{\varphi_0}{f_0(l,k)} \int_x^\infty cV \varphi_0 f_0 \varphi_2 dx' \right). \tag{9.31}$$

9.3   Connection of Potential Representation Method with Green's Functions

In this section, we will observe that the integral equations (9.30) and (9.31) obtained for the Schrödinger equation using potential representation can be obtained using Green's functions. In order to achieve this, we can combine the integrals in Eqs. (9.30) and (9.31) in the following manner:

$$\varphi_1(l,k,x) = \frac{1}{f_0(l,k,x)}$$

$$\times \left(f_0(l,k,x)\alpha - \frac{1}{f_0(l,k)} \times \int_0^\infty f_0(l,k,x_\rangle)\varphi_0(l,k,x_\langle) \right.$$

$$\left. \times cV(x')f_0(l,k,x')\varphi_1(l,k,x')dx' \right), \tag{9.32}$$

$$\varphi_2(l,k,x) = \frac{1}{\varphi_0(l,k,x)}$$

$$\times \left(\varphi_0(l,k,x)\beta - \frac{1}{f_0(l,k)} \times \int_0^\infty f_0(l,k,x_\rangle)\varphi_0(l,k,x_\langle) \right.$$

$$\left. \times cV(x')\varphi_0(l,k,x')\varphi_2(l,k,x')dx' \right), \tag{9.33}$$

Here, $x_\rangle$ is greater than x, x' and $x_\langle$ is less than x, x'.

According to [3] the kernel of integral equations (9.32) and (9.33),

$$G(x_\rangle, x_\langle) = -\frac{f_0(l, k, x_\rangle)\varphi_0(l, k, x_\langle)}{f_0(l, k)} \tag{9.34}$$

makes an appearance in Green's function without using the standard condition

$$\hat{L}G(x, x') = \delta(x - x'). \tag{9.35}$$

The proposed method of potential representation makes it possible to obtain Green's functions like production of linearly independent unperturbed solutions for the boundary conditions presented.

For the potentials with a long interaction radius $V_0(x) = \frac{1}{x}$, instead of the functions f_0 and φ_0, in the spherical Coulomb functions (9.34) must be used [2] when we want to estimate the modified Coulomb phase shifts by a short-radius potential $V(x)$ guaranteeing the descent of the integrals (9.32) and (9.33). For the free movement, we have [1]

$$f_0(l, k, x) = \left(\frac{\pi k x}{2}\right)^{\frac{1}{2}} \exp\left\{-\frac{i\pi(l+1)}{2}\right\} H_{l+\frac{1}{2}}^{(2)}(kx), \tag{9.36}$$

$$\varphi_0(l, k, x) = 2^{l+\frac{1}{2}} k^{-l-\frac{1}{2}} \Gamma\left(l + \frac{3}{2}\right) x^{\frac{1}{2}} J_{l+\frac{1}{2}}^{(1)}(kx), \tag{9.37}$$

$$f_0(l, k) = 2^{l+\frac{1}{2}} \sqrt{\frac{2}{\pi}} \Gamma\left(l + \frac{3}{2}\right) k^{-1} \exp\left\{-\frac{i\pi l}{2}\right\}, \tag{9.38}$$

where $H_{l+\frac{1}{2}}^{(2)}(kx)$ are the Hankel functions of the second kind and $J_{l+\frac{1}{2}}^{(1)}(kx)$ are the Bessel functions of the first sort.

Using the relation between special functions [4] and both Bessel and Hankel spherical functions, we obtain a coincidence between Green's function (9.34) and the function obtained by a standard method [3].

9.4   The Scattering Matrix

The scattering matrix can be expressed by the solutions (9.30) and (9.32) of the radial equation (9.5) by an alternate method as follows:

$$S_l(k) = \frac{\varphi_1(l, k, 0) \lim_{x \to 0} x^l f_0(l, k, x)}{\varphi_1(l, -k, 0) \lim_{x \to 0} x^l f_0(l, -k, x)} \exp\{i\pi l\}. \tag{9.39}$$

After substituting (9.30) in the last formula, the scattering matrix can be obtained

$$S_l(k) = \frac{1 + \frac{1}{f_0(l,k)} \int_0^\infty cV(x') f_0(l, k, x') \varphi_0(l, k, x') \varphi_1(l, k, x') dx'}{1 + \frac{1}{f_0(l,-k)} \int_0^\infty cV(x') f_0(l, -k, x') \varphi_0(l, k, x') \varphi_1(l, -k, x') dx'}$$
$$\cdot S_l^0(k) \tag{9.40}$$

$$S_l^0(k) = \frac{f_0(l, k)}{f_0(l, -k)} \exp(i\pi l). \tag{9.41}$$

The obtained formula can be applied for the scattering matrix where functions $f_0, \varphi_0, f_0(l, k)$ are defined by (9.36)–(9.38). In this case, after solving the integral equation (3.1) we can obtain the scattering matrix $S_l(k)$. In a common case, when $V_1(x) \neq 0$ we can use (9.40) and (9.41) for the separation of the scattering matrix $S_l^0(k)$ for potential $V_1(x)$ from $S_l(k)$ for the sum of the potentials $V_1(x)$ and $V(x)$:

$$S_l(k) = \frac{f_0(l, k) + \int_0^\infty cV(x') f_0(l, k, x') \varphi_0(l, k, x') \varphi_1(l, k, x') dx'}{f_0(l, k) + S_l^0(k) \exp\{-i\pi l\} \int_0^\infty cV(x')}$$
$$\frac{}{f_0(l, -k, x') \varphi_0(l, k, x') \varphi_1(l, -k, x') dx'}$$
$$\times S_l^0(k). \tag{9.42}$$

The scattering matrix $S_l(k)$ calculated by using the method of potential representation (9.42), (9.32) or obtained from measurements when the additional potential $V(x)$ is included gives a possibility to allot the scattering matrix $S_l^0(k)$ for the previously exposed potential $V_1(x)$. This method can be very useful for investigating the scattering dependence on the particle energies and interactions with different nuclei.

The expression obtained for the scattering matrix and the integral equations (9.32), (9.33) allow to separate the well-known potential and scattering matrix (9.41) for the investigation of a new additional potential $V(x)$ and additional effects using scattering experiments and the formulas obtained.

References

1. de Alfaro, V. and Regge, T. (1965). *Potential Scattering*, North-Holland Publishing Company, Amsterdam, p. 274.
2. Guter, R. S. and Fuller, A. R. (1976). *Differential equations*, High School, Moscow, p. 304.
3. Byron, F. W. and Fuller, R. W. (1992). *Mathematics of Classical and Quantum Physics*, Dover Publications, New York, p. 661.
4. Korn, G. A. and Korn, T. M. (1961). *Mathematical Handbook for Scientists and Engineers*, McGraw-Hill, USA, p. 1097.

Chapter 10

The Separation of the Scattering Matrix from the Coulomb Field

10.1 Introduction

Using the modified method of undetermined coefficients, integral equations are found for the solution of the Schrödinger equation in potential representation for the short-range and Coulomb potentials. The method is based on the multiplicative perturbation theory. Here, we obtained the relation between the scattering matrix for short-range acting forces and the scattering matrix for simultaneously acting Coulomb potential A[5].

We will consider the common properties of solutions for the radial Schrödinger equation

$$\frac{d^2 f}{dx^2} + (k^2 - cV(x) - V_0(x))f = 0, \qquad (10.1)$$

where

$$c = \frac{2m}{\hbar^2}, \quad V_0(x) = cV_1(x) + \frac{l(l+1)}{x^2}, \quad V_1(x) = Z_1 Z_2 \frac{e^2}{x}. \qquad (10.2)$$

The potential $V_1(x)$ at zero has asymptotic behavior at $\frac{1}{x^\alpha}$, $\alpha < 2$, and at infinity, it is $\frac{1}{x^{1+\varepsilon}}$, $\varepsilon > 0$, $\varepsilon \neq 1$. According to the potential representation method [2], the solutions (10.1) for a continuous spectrum can be represented as follows:

$$f(l, k, x) = \varphi(l, k, x) f_0(l, k, x). \qquad (10.3)$$

Here, $\varphi(l, k, x)$ depends on the potential $V(x)$, and $f_0(l, k, x)$ is a solution of the equation

$$f_0'' + \left(k^2 - c\frac{Z_1 Z_2 e^2}{x} - \frac{l(l+1)}{x^2} \right) f_0 = 0. \tag{10.4}$$

Substituting (10.3) into (10.1) and by considering (10.4), we obtain the equation for function φ as follows:

$$f_0' \varphi'' + 2f_0' \varphi' - cV f_0 \varphi = 0. \tag{10.5}$$

10.2   Obtaining Integral Equations

We will investigate the solutions of the Eq. (10.1) according to the method of undetermined coefficients [1] and we will be using the following equation:

$$f^\pm = C_1(x)F_l + C_2(x)U_l^\pm, \tag{10.6}$$

where F_l and $U_l^\pm$ are spherical Coulomb functions [2]

$$
\begin{aligned}
F_l(\gamma, kx) &= Ce^{ikx}(kx)^{l+1}F(l+1+\gamma i, 2l+2, -2ikx), \\
U_l^\pm(\gamma, kx) &= \mp 2ie^{\mp i\sigma_l}C_l e^{\pm ikx}(kx)^{l+1} \\
&\quad \times W_1(l+1\pm\gamma i, 2l+2, \mp 2ikx),
\end{aligned} \tag{10.7}
$$

$$\gamma = \frac{Z_1 Z_2 e^2}{\hbar v}, \quad C_l = 2^l e^{-\frac{\pi}{2}\gamma}\frac{|\Gamma(l+1+i\gamma)|}{(2l+1)!},$$

$$\sigma_1 = \arg\Gamma(l+1+i\gamma). \tag{10.8}$$

These functions satisfy the following boundary conditions [2, 3]:

$$F_l \underset{x\to\infty}{\approx} \sin\left(kx - \gamma\ln 2kx - \frac{\pi \cdot l}{2} + \sigma_l \right), \tag{10.9}$$

$$U_l^\pm \underset{x\to\infty}{\approx} \exp\left(\pm i\left(kx - \gamma\ln 2kx - \frac{\pi \cdot l}{2} \right) \right)\bigg\}, \tag{10.10}$$

$$F_l \underset{x\to 0}{\approx} C_l(kx)^{l+1}, \tag{10.11}$$

$$U_l^\pm \underset{x\to 0}{\approx} \frac{e^{\mp i\sigma_l}}{(2l+1)C_l}(kx)^{-1}. \tag{10.12}$$

The first partial solution of (10.5) is presented in the following way:

$$\varphi_1(l, \pm k, x) = C_1(x)\frac{F_l(\gamma, kx)}{U_l^{\pm}(\gamma, kx)} + C_2(x) \tag{10.13}$$

with the following boundary condition:

$$\lim_{x \to \infty} \varphi_1(l, \pm k, x) = 1, \tag{10.14}$$

and the second solution is

$$\varphi_2(l, \pm k, x) = C_1(x)\frac{U_l^{\pm}(\gamma, kx)}{F_l(\gamma, kx)}C_2(x) \tag{10.15}$$

under the boundary condition

$$\lim_{x \to 0} \varphi_2(l, k, x) = 1. \tag{10.16}$$

Subsequently, substituting (10.13) and (10.15) into (10.5) and considering the boundary conditions (10.14), (10.16) and the Wronskian

$$U_l^{\pm}\frac{dF_l}{dx} - F_l\frac{dU_l^{\pm}}{dx} = e^{\mp i\sigma_l}k, \tag{10.17}$$

we obtain

$$\varphi_1^{\pm} = \alpha(\pm k) - \frac{c}{k}e^{\pm i\sigma_l}\int_0^x VU_l^{\pm}F_l\varphi_1^{\pm}dx'$$
$$- \frac{c}{k}e^{\pm i\sigma_l}\frac{F_l}{U_l^{\pm}}\int_x^{\infty} V(U_l^{\pm})^2\varphi_1^{\pm}dx', \tag{10.18}$$

$$\alpha(\pm k) = 1 + \frac{c}{k}e^{\pm i\sigma_l}\int_0^{\infty} VU_l^{\pm}F_l\varphi_1^{\pm}dx',$$

$$\varphi_2^{\pm} = \beta(\pm k) - \frac{c}{k}e^{\pm i\sigma_l}\frac{U_l^{\pm}}{F_l}\int_0^x V(F_l)^2\varphi_2^{\pm}dx'$$
$$- \frac{c}{k}e^{\pm i\sigma_l}\int_x^{\infty} VF_lU_l^{\pm}\varphi_2dx', \tag{10.19}$$

$$\beta(\pm k) = 1 + \frac{c}{k}e^{\pm i\sigma_l}\int_0^{\infty} VU_l^{\pm}F_l\varphi_2^{\pm}dx'.$$

Here, we mark $\varphi(l, \pm k, x) = \varphi^{\pm}$. Integrating (10.18) and (10.19) on the right-hand side, we obtain

$$\varphi_1^{\pm} = \frac{1}{U_l^{\pm}} \left(U_l^{\pm} \alpha(\pm k) + \int_0^{\infty} G_c^{\pm} V U_l^{\pm} \varphi_l^{\pm} dx' \right), \qquad (10.20)$$

$$\varphi_2^{\pm} = \frac{1}{F_l} \left(F_l \beta(\pm k) + \int_0^{\infty} G_c^{\pm} V F_l \varphi_2^{\pm} dx' \right). \qquad (10.21)$$

According to A[6], $G^{\pm}$ represents Green's functions for scattering task

$$G^{\pm}(x_{\langle}, x_{\rangle}) = -e^{\pm i\sigma_l} \frac{F_l(\gamma, kx_{\langle}) U_l^{\pm}(\gamma, kx_{\rangle})}{k}. \qquad (10.22)$$

The integrals that have been presented and integral equations that have been obtained are defined by fast decreasing $V(x)$ at infinity.

Verifying the correctness of the obtained solutions $\varphi^{\pm}$, expressed by the integral equations (10.18), we obtain

$$\frac{d}{dx} \varphi_l^{\pm} = -\frac{1}{U_l^2} \int_x^{\infty} cV (U_l^{\pm})^2 \varphi_l^{\pm} dx', \qquad (10.23)$$

$$\frac{d^2}{dx^2} \varphi_l^{\pm} = 2 - \frac{(U_l^{\pm})'}{U_l^3} \int_x^{\infty} cV (U_l^{\pm})^2 \varphi_1^{\pm} dx' + cV \varphi_1^{\pm}. \qquad (10.24)$$

Substituting (10.23) and (10.24) in (10.5) and keeping in mind the fact that $f_0 = U_l^{\pm}$, we obtain $\varphi_l^{\pm}$, which represents the solution of the Schrödinger equation in potential representation satisfying the boundary condition (10.14). Similarly, we can obtain $\varphi_2^{\pm}$ as well.

The obtained integral equations (10.18) and (10.19) are integral equations of the second type of Volterra equations [4] if $\alpha(\pm k)$, $\beta(\pm k)$ can be represented by two integrals in the intervals 0, x and x, ∞. Then, from (10.18) and (10.19), we can obtain

$$\varphi_1^{\pm} = 1 + \int_0^{\infty} K_1^{\pm}(x, x') \varphi_1^{\pm}(l, k, x') dx', \qquad (10.25)$$

$$K_1^{\pm}(x, x') = \frac{c}{k} e^{\pm i\sigma_l} \left(F_l(\gamma, kx') - \frac{F_l(\gamma, kx)}{U_l^{\pm}(\gamma, kx)} U_l^{\pm}(\gamma, kx') \right)$$
$$\times V(x') U_l^{\pm}(\gamma, kx'), \qquad (10.26)$$

$$\varphi_2^{\pm} = 1 + \int_0^x K_2^{\pm}(x, x') \varphi_2^{\pm}(l, k, x') dx', \qquad (10.27)$$

$$K_2^{\pm}(x, x') = \frac{c}{k} e^{\pm i\sigma_l} \left(U_l^{\pm}(\gamma, kx') - \frac{U_l^{\pm}(\gamma, kx)}{F_l(\gamma, kx)} F_l(\gamma, kx') \right)$$
$$\times V(x') F_l(\gamma, kx'). \qquad (10.28)$$

Equations (10.25) and (10.27) have unique solutions [5] only when the kernels $K_1^{\pm}$ and $K_2^{\pm}$ are continuous in the intervals $0 \leq x \leq x' < \infty$ and $0 \leq x' \leq x < \infty$. When $U_l^{\pm}$ for any x are not equal to zero and when F_l are limited, then the functions $K_1^{\pm}$ (10.28) are continuous. Also, poles of the kernels $K_2^{\pm}$ coincide with zeros of F_l, i.e.,

$$\varphi_2^{\pm} F_l = F_l(\gamma, kx) + \int_0^x N_2^{\pm}(x, x') \varphi_2^{\pm}(l, k, x') F_l(\gamma, kx') dx',$$
$$(10.29)$$

where the kernel

$$N_2^{\pm}(x, x') = F_l(\gamma, kx) K_2^{\pm}(x, x'), \qquad (10.30)$$

according to (10.28) in the interval $0 \leq x' \leq x < \infty$ is a continuous function that guarantees continuity and uniqueness of solutions $\varphi_2^{\pm} F_l$.

The integral equations (10.25) and (10.29) that are obtained can be solved by the iteration method with initial values $\varphi_{1,0}^{\pm} = \varphi_{2,0}^{\pm} = 1$. Convergence of solution depends on the potential decreasing velocity at infinite distance. The changes of solutions for the Schrödinger equation when the potential $V(x)$ is added are introduced as multipliers $\varphi_1^{\pm}$ and $\varphi_2^{\pm}$. Keeping in mind the fact that we had not estimated the smallness of the perturbation potential, the multiplicative or potential representation method is the most preferred method with the Rayleigh–Schrödinger perturbation theory [6]. From (10.29) and (10.30), it follows that zeros $\varphi_2^{\pm} F_l$ are removed from the position in function F_l as a result of the interaction potential $V(x)$. Using the equations obtained, we can investigate the changes in the analytical properties of solutions connected with the concrete potential.

10.3   Obtaining the Scattering Matrix

The scattering matrix for the short-radius potential can be obtained from [4] and (10.18) in the following way A[6]:

$$e^{2i\delta_l(k)} = \lim_{x \to 0} \frac{\varphi_1(l, -k, x)U_l^-(\gamma, kx)}{\varphi_1(l, k, x)U_l^+(\gamma, kx)} e^{-2i\sigma_l}. \qquad (10.31)$$

After calculating the limits for pure short-radius potential, we obtain

$$e^{2i\delta_l(k)} = \lim_{x \to 0} \frac{\varphi_1(l, -k, x)}{\varphi_1(l, k, x)} = \frac{\alpha(-k)}{\alpha(k)}. \qquad (10.32)$$

To verify the obtained formula, we can use the Wronskian theorem [2] for linear differential equations

$$\lim_{x \to \infty} W(F_l, \varphi_2 F_l) = c \int_0^\infty V F_l^2 \varphi_2 dx'. \qquad (10.33)$$

From the Wronskian theorem for ordinary differential equations of the second order [4], it follows that any solution [5] of the Schrödinger equation can be expressed by two linearly independent solutions

$$\varphi_2 F_l = \frac{f(l, k)\varphi_1^+ U_l^+ - f(l, -k)\varphi_1^- U_l^-}{2ik}. \qquad (10.34)$$

Here, the Jost functions are expressed by the Wronskians

$$f(l, k) = W(\varphi_1 U_l^-, \varphi_2 F_l), \qquad (10.35)$$

$$f(l, -k) = W(\varphi_1 U_l^+, \varphi_2 F_l). \qquad (10.36)$$

When the short-radius potential $V(x)$ is identically equal to zero, from Eq. (10.17) we obtain

$$f_0(l, k) = ke^{i\sigma_l}, \qquad (10.37)$$

$$f_0(l, -k) = ke^{-i\sigma_l}. \qquad (10.38)$$

From the boundary condition $\lim_{x \to 0} \varphi_2 F_l = 0$ and Eqs. (10.35)–(10.38), we have A[6]

$$e^{2i(\delta_l + \sigma_l)} = \frac{f(l, k)}{f(l, -k)}. \qquad (10.39)$$

Using expression (10.39) and substituting (10.9) and (10.10) into (10.33), we can get (10.39). The obtained formula (10.39) can be used for calculating the phase shift δ_l for the short-radius potential when Coulomb forces are acting. We generalize the Schwinger method [6], which can be used for calculations when Coulomb interaction is absent. Expression of the scattering matrix (10.32), (10.25) allows in the separation of the scattering matrix for short-radius potentials, which can be investigated separately by scattering experiments from the well-known Coulomb potentials. For practical calculations of phase shifts, integral equations (10.25) are more handy than (10.20) because integration must be provided in a more tight interval. When we refuse factorization (10.3), the proposed method shall transform to wave distortions.

References

1. Guter, R. S. and Janpolski, A. R. (1976). *Potential Scattering*, Vysheya Shkola, Moscow, p. 304 (in Russian).
2. Messia, A. (1978). *Quantum Mechanics*, Vol. 1, Science, Moscow, p. 477.
3. Byron, F. W. and Fuller, R. W. (1992). *Mathematics of Classical and Quantum Physics*, Dover Publications, New York, p. 661.
4. Baz, A. I., Zeldowich, J. B., and Perelomov, A. M. (1966). *Scattering, Reactions and Decay in Nonrelativistic Quantum Mechanics*, Science, Moscow, p. 339 (in Russian).
5. Kamke, E. (1971). *Handbook of Differential Equations*, Academic Press, Boston, p. 576.
6. Wu, Yu. T. and Omura, T. (1969). *Quantum Theory of Scattering*, Nauka, Moscow, p. 451 (in Russian).

Chapter 11

The General Solution for Bound States of the Woods–Saxon Potential

11.1 Introduction

The Schrödinger equation for negative energies in the potential representation A[9] was considered for transforming it to integral equations. We estimated the wave function by multiplying with the known solution of the function, which depends on additional interaction potential. We considered integral equations for defining the eigenfunctions and eigenvalues of the perturbed system using unperturbed solutions of the Schrödinger equation [1–3]. We applied the integral equations obtained for finding the eigenfunctions and eigenvalues for Woods–Saxon potential using a harmonic oscillator such as the model potential. Usually, we consider nuclei as non-relativistic systems. But the nuclear forces have a repulsive core and a great spin–orbit interaction of relativistic origin, thus changing A[14] kinetic and potential energies. The semi-relativistic calculations [4] can be provided using solutions of non-relativistic Schrödinger equation. In Ref. A[1], we present the analytical solution of the Schrödinger equation for the Woods–Saxon potential only when the orbital moment is equal to zero. The presented solution cannot be applied for the real physical problems, where the term $\frac{l(l+1)}{r^2}$ must be included into the radial Schrödinger equation. The exact analytical solution in this case for the Woods–Saxon potential has never been obtained. We can obtain eigenvalues and wave functions of the radial Schrödinger equation solving the system of the integral equations A[9, 17, 19]. The exact solutions were obtained at large distances and therefore can be used for relativistic corrections for the mass of nucleus or kinetic energy. All the theoretical calculations with the spherically symmetric Woods–Saxon potentials we provided for investigations of the heavy nuclei use the shell model approach. We compared calculations with the experimental energy levels of ^{208}Pb A[14]. The method of general solution was applied as the original method for finding the wave functions and the eigenvalues of the Schrödinger equation. The proposed methods can be very useful and relevant for considering the shell model applications for superheavy nuclei A[17, 18]. Solving the integral equations is always a more exact operation than solving differential equations by the finite-difference approximation. In this chapter, some modified potential representation method A[4] for negative energies is applied. The main idea of this method is an expression of the radial wave function via the unperturbed wave function multiplied with the function that depends on the perturbation potential. We shall solve the Schrödinger equation and obtain the neutron energy levels E_{nLj} of the ^{208}Pb nucleus for the Woods–Saxon potential [4, 7]:

$$V_s(r) = -\frac{V_n}{1 + e^{\alpha(r-R)}}, \tag{11.1}$$

and the spin–orbit potentials

$$(\vec{s}\,\vec{l})_{j1} = \frac{1}{2}l\hbar^2, \quad j1 = L + \frac{1}{2}, \tag{11.2}$$

$$(\vec{s}\cdot\vec{l})_{j2_1} = -\frac{1}{2}(L+1), \quad j2 = L - \frac{1}{2}, \tag{11.3}$$

for total angular moments $j1$ and $j2$. Here, we used the energy units MeV (mega electron volts) and distance ($fm = 10^{-15}$ m). The Schrödinger equation with potential (11.1) cannot be solved analytically when $L \neq 0$ without approximate expansions and evaluation of accuracy [7]. In this case, comparing these results is not expedient. Usually, solutions for the potentials (11.1) and (11.2) can be found using discretization A[5] or Green's functions methods [1].

In earlier papers, the potential representation method for positive and negative energies has been derived, where the perturbed solution can be obtained by multiplying the unperturbed solution by a multiplier depending on the perturbation potentials A[5, 9]. This method is equivalent to the modified A[5] method of unidentified coefficients [3], A[10, 15], i.e.,

$$U_\alpha = C_{1\alpha}(r)F_{nL}(r) + C_{2\alpha}(r)U_{nL}(r). \tag{11.4}$$

Here, the perturbed solutions are expressed by a linear combination of nonphysical $E_{nL}(r)$, $(\lim_{r\to\infty} F_{nL}(r) \to \infty)$ and physical $U_{nL}(r)$, $(\lim_{r\to\infty} U_{nL}(r) \to 0)$ solutions for unperturbed A[9] linearly independent solutions. Using this method in modified form

$$U_\alpha = \left[C_{1\alpha}(r)\frac{F_{nL}(r)}{U_{nL}(r)} + C_{2\alpha}(r) \right] U_{nL}(r) = \varphi_\alpha(r)U_{nL}(r), \tag{11.5}$$

we obtained the integral equations and Green's functions A[9] for finding the eigenfunctions and eigenvalues of the Schrödinger equation for Woods–Saxon potential.

11.2 The Derivation of Integral Equations

Here, we consider the derivation of integral equations and convergence of solutions when the perturbed solution in (11.4) substituting $C_{1\alpha}(r) = 0$ was presented. A general solution of the Schrödinger equation in the potential representation was put forward in the form of integral equations. The radial eigenfunctions in potential representation of the Schrödinger equation A[4] can be expressed in the form $U_\alpha(r) = \varphi_\alpha(r)U_{nL}(r)$ as a product of the unperturbed physical wave function $U(r)_{nL}$ and multiplier $\varphi_\alpha(r)$ depending on the perturbation potential $V_\delta(r)$. Solving the Schrödinger equation

$$\frac{d^2}{dr^2}U_\alpha - \frac{L(L+1)}{r^2}U_\alpha + C(E_\alpha - V_0(r) - V_\delta(r))U_\alpha = 0,$$

$$V_\delta(r) = V_S(r) - V_0(r), \quad C = \frac{2m}{\hbar^2}, \tag{11.6}$$

for the Woods–Saxon potential $V_S(r)$, we can use the model potential $V_0(r)$ and the corresponding analytical solutions $U_{nL}(r)$. If we use the

Woods–Saxon potential (11.1) and the spin–orbit potential (11.2), the Schrödinger equation cannot be solved analytically. In this case [7], we can use the radial harmonic oscillator potential

$$V(r) = \frac{m\omega^2 r^2}{2} - V_0, \quad w = d w_0, \quad w_0 = 41 \cdot A^{-1/3} \frac{\text{MeV}}{\hbar} \quad (11.7)$$

as the model potential with parameter $d (d \in [0.6, 1.25])$ and eigenfunctions

$$U_{nL} = e^{-0.5\rho} \rho^{0.5(L+1)} \sum_{k=0}^{n-1} a_k \rho^k, \quad \rho = m w r^2 / \hbar, \quad (11.8)$$

$$a_{k+1} = \frac{k - 1/2(\varepsilon_{nL} - L - 3/2)}{(k+1)(k+L+3/2)} a_k, \quad a_0 = 1, \quad (11.9)$$

with known eigenvalues

$$E_{nL} = \varepsilon_{nL} \hbar \omega - V_n, \quad \varepsilon_{nL} = 2n + L - \frac{1}{2}, \quad (11.10)$$

as unperturbed or model solutions A[4]. The varying parameter $d \in [0.6, 1.2]$ was introduced for the optimization of the model potential (11.7) and faster convergence of the solutions (11.8). The asymptotic behavior of model wave functions satisfies the following boundary conditions:

$$\lim_{r \to 0} U_{nL} = \left(r^2 \cdot \frac{m\omega}{\hbar} \right)^{0.5(L+1)} \to 0,$$

$$\lim_{r \to \infty} U_{nL} = a_{n-1} e^{-0.5\rho} \left(r^2 \cdot \frac{m\omega}{\hbar} \right)^{0.5(L+1)+n-1} \to 0. \quad (11.11)$$

The radial eigenfunctions $U_\alpha(r)$ in general representation A[4] for the negative energies can be expressed as a product of the unperturbed wave function $U_{nL}(r)$ and multiplier $\varphi_\alpha(r)$ in the case where the following boundary conditions are satisfied:

$$U_\alpha(r) = \varphi_\alpha(r) U_{nL}(r), \quad 0 < r < \infty,$$

$$\lim_{r \to 0} U_\alpha(r) \to 0, \quad \lim_{r \to \infty} U_\alpha(r) \to 0. \quad (11.12)$$

This representation was considered because it can be applied for positive energies as well A[4].

Theorem 1. *The general representation* (11.11) *for the negative energies or bound states can be used only for principal quantum number* $n = 1$, *where the nonperturbed solutions* $U_\alpha(r)$ *have zeros only at initial point and at infinity.*

Proof. Eigenfunctions $U_\alpha(r)$ of Eq. (11.6) are expressed as a product of eigenfunctions $U_{nL}(r)$ of unperturbed Eq. (11.6), when the function $\varphi_\alpha(r)$ depends on the perturbation $V_\delta(r)$. The eigenfunctions $U_{nL}(r)$ are polynomials of $n - 1$ degree and, in the case $n > 1$, we change the boundary conditions (11.11) introducing $n - 1$ zeros. The boundary conditions are not changed only for $n = 1$ or using the method (11.9) of unidentified coefficients [3], A[10, 15]. The representation of solution using the method of unidentified coefficients (11.9) is not equivalent to the representation (11.12) when the solutions $U_{nL}(r)$ have zeros corresponding to the boundary conditions (11.12) and also within the region of integration interval. In the case $n \geq 1$, simple multiplier $\varphi_\alpha(r)$ in (11.12) cannot change the position of zeros and can represent the perturbed solution only for $n = 1$. The Theorem 1 is thus proved.

We have proved this postulate [1, 3], A[4] for the presented theorem using the method of indefinite coefficients. From this theorem, the connection between the solutions for short-range potentials and Coulomb potentials has been obtained for negative A[15] and positive A[5] energies. In potential representation [1, 3], A[15], the perturbed radial function, $U_{\alpha nL}(r)$ with the set of quantum numbers αnL can be expressed as a product of unperturbed wave function U_{nL} with the set of quantum numbers nL, $U_{\alpha nL}(r) = \varphi_\alpha(r) \cdot U_{nL}(r)$ on the multiplier $\varphi_\alpha(r)$ depending on the perturbation potential $V_\delta(r)$. The main problem considered in this chapter is the interpretation and the convergence of the calculated eigenvalues using the harmonic oscillator potential for the modeling of the Woods–Saxon potential. At large distances, these potentials have essentially different asymptotical behavior, i.e., r^2 and $e^{-\alpha r}$, $\alpha > 0$, respectively. The negative eigenvalues of energies $E_\alpha = E_{nL} + \Delta E_{nLj}$ for different meanings of quantum numbers n, L, j, including the spin–orbit interactions [5, 7], can be expressed by the eigenvalues E_{nL} for model potential (11.7). Here, substituting $U_\alpha(r)$, we have the equation

$$\frac{d}{dr}\left(U_{nL}^2 \frac{d}{dr}\varphi_\alpha\right) = -CU_{nL}(\Delta E_{nL\alpha} - V_\delta(r))\varphi_\alpha U_{nL}, \qquad (11.13)$$

which can be transformed to the following form:

$$\frac{d}{dr}\left(\frac{d}{dr}U_{nL}^2\varphi_\alpha - \varphi_\alpha \frac{d}{dr}U_{nL}^2\right) = -CU_{nL}(\Delta E_{nL\alpha} - V_\delta(r))\varphi_\alpha U_{nL}.$$

$$(11.14)$$

Upon integrating, we obtain the expression

$$\frac{d}{dr}U_{nL}^2\varphi_\alpha - \varphi_\alpha \frac{d}{dr}U_{nL}^2 = -\int_0^r CU_{nL}(\Delta E_{nL} - V_\delta(r))\varphi_\alpha U_{nL}dr$$

$$(11.15)$$

where substituting (11.13) $U_\alpha(r)$ in the last formula, we find

$$\frac{dU_\alpha}{dr} = \frac{U_\alpha}{U_{nL}}\frac{dU_{nL}}{dr} - \frac{C}{U_{nL}}\int_0^r U_{nL}(\Delta E_{nL\alpha} - V_\delta)U_\alpha dz. \qquad (11.16)$$

Considering the boundary condition (11.11) when $r \to 0$ and integrating, we obtain the integral equation

$$U_\alpha = \int_0^r \frac{U_\alpha}{U_{nL}}\frac{dU_{nL}}{dr}dy - \int_0^r \frac{Cdy}{U_{nL}}\int_0^y U_{nL}(\Delta E_{nL\alpha} - V_\delta)U_\alpha dz.$$

$$(11.17)$$

Introducing the following dimensionless constant for the nucleons in heavy nucleus

$$\lambda = C \cdot \text{MeV} = 0.0486, \qquad (11.18)$$

we will represent energies in the mega electron volt (MeV) units. Substituting the approximate expressions of perturbed eigenfunctions

$$U_{\alpha N} = \sum_{m=0}^N \lambda^m \varphi_{m\alpha} U_{nL}, \quad \varphi_{0\alpha} = 1, \qquad (11.19)$$

into the integral equation (11.17) and solving by the iteration method, we obtain for $N + 1$ iteration step

$$U_{\alpha N+1} = \int_0^r \frac{U_{\alpha N}}{U_{nL}} \frac{dU_{nL}}{dr} dy - \lambda \int_0^r \frac{dy}{U_{nL}}$$

$$\times \int_0^y U_{nL}(\Delta E_{nL\alpha N} - V_\delta)U_{\alpha N}dz \quad N = 0, 1, 2, \ldots .$$

$$(11.20)$$

The presented approximate perturbed eigenfunctions must satisfy the same boundary conditions like unperturbed those for the eigenfunctions (11.11) according to Theorem 1. Requiring that $\lim_{r \to \infty} U_{\alpha N}(r) \to 0$, from (11.17) we obtain the meaning of perturbed eigenvalues

$$\Delta E_{nLj} = \Delta E_{nL\alpha} = \frac{\int_0^\infty U_{nL}V_\delta \varphi_\alpha U_{nL}dr}{\int_0^\infty U_{nL}\varphi_\alpha U_{nL}dr}. \quad (11.21)$$

If after every iteration, we use normalized approximate solutions

$$\int_0^\infty U_{nL}U_{\alpha N}dr = 1, \quad (11.22)$$

then the perturbations ΔE_{nLjN} of energy E_{nL} can be expressed substituting the polynomial expansion (11.19) into (11.20)

$$\Delta E_{nLjN} = \int_0^\infty U_{nL}V_\delta(r)U_{nL}dr + \lambda \int_0^\infty U_{nL}V_\partial(r)\varphi_{1\alpha}U_{nL}dr$$

$$+ \cdots + \lambda^N \int_0^\infty U_{nL}V_\partial(r)\varphi_{N\alpha}U_{nL}dr. \quad (11.23)$$

This condition is satisfied if the function U_{nL} is continuously bounded and both functions, U_{nL} and $U_{\alpha N}$, satisfy the same boundary conditions (11.12) at the origin and similarly if (11.23) $\lim_{r \to \infty} U_{\alpha N}(r) \to 0$ at the infinity.

11.3 The Accuracy and Convergence of the Obtained Solutions

At first, we calculated the one-neutron energy levels nLj of ^{208}Pb for the case of Woods–Saxon potential A[15] for the wave functions

Table 11.1.   The energy levels $\bar{E}_{nLJ}$ of the neutrons that
are presented in Ref. A[14] and $E_{nLj} = E_{nL} + \Delta E_{nLj}$ of
the nucleus ^{208}Pb calculated by (11.10) and (11.21) using
the model potential (11.7) for Woods–Saxon potential with
parameters $V_n = 44.0\,\mathrm{Mev}$, $R = 7.6\,\mathrm{fm}$, $\alpha = 1.512\,\mathrm{fm}^{-1}$,
$\kappa = 0.353\,\mathrm{fm}^2$, number of integration steps 200 at maximum
integration distance 20.48 fm with partial accuracy 0.0001 and
iterations number N.

nLj	$\bar{E}_{nLj}\,\mathrm{MeV}$	$E_{nLj}\,\mathrm{MeV}$	d	N
$1i_{13/2}$	-8.617	-8.436	1.04	4
$1h_{9/2}$	-10.93	-10.86	1.08	6
$1h_{11/2}$	-14.93	-14.82	0.98	5
$1f_{5/2}$	-24.84	-24.75	0.92	5
$1f_{7/2}$	-26.53	-26.40	0.86	6
$1s_{1/2}$	-39.97	-39.83	0.62	4

$U_\alpha(r)$ without zeros in the region $r \in (0, \infty)$. Here, $n = 1, 2, 3, \ldots$ are
the principal quantum numbers, $L = 0, 1, 2, 3, 4, 5, \ldots$ are the orbital
quantum numbers marked as $s, p, d, f, g, h, \ldots$, and j is the total
angular momentum. The obtained energy levels or eigenvalues E_{nLj}
for states with the principal quantum numbers $n = 1$ are compared
with results $\bar{E}_{nLj}$ calculated by the discretization method A[14] and
presented in Table 11.1. The obtained results are in good coincidence
when $n = 1$. For relative accuracy 0.0001 of calculated energy levels,
the iteration numbers N are small.

We chose the maximum integration region for (11.20)–(11.22)
$R_m = 16.68\,\mathrm{fm}$. We provided similar calculations A[14] with 14 fm
only. For greater R_m (20.32 fm), we obtained the same number of
iterations N and the same energy levels or eigenvalues E_{nlj}.

From Table 11.2, we obtain that when model potentials are
optimized (11.7) by changing the parameter d, the eigenvalues E_{nLj}
are defined more exactly for parameters used for the potentials.
When the perturbation ΔE_{nLj1} is added to unperturbed energy
E_{nL}, we obtain the first approach of energy. The following values
of energy $E_{nLjN} = E_{nL} + \Delta E_{nLjN}$ after five iterations practically
coincide. Comparing the results presented in Tables 11.1 and 11.2,
we obtain that for optimal model potentials (11.7) the eigenvalues

Table 11.2.   The energy levels $\bar{E}_{nLj}$ of the neutrons [6] and $E_{nLj} = E_{nL} + \Delta E_{nLj}$ of the ^{208}Pb nucleus calculated by (11.10), (11.21) using the model potential (11.7) for Woods–Saxon potential with parameters $V_n = 45.7\,\mathrm{Mev}$, $R = 7.6\,\mathrm{fm}$, $\alpha = 1.538\,\mathrm{fm}^{-1}$, $\kappa = 0.374\,\mathrm{fm}^2$. The number of integration steps 200 at maximum distance 20.48 fm with partial accuracy 0.0001 and iterations number N.

nLj	$\bar{E}_{nLj}$ MeV	E_{nLj} MeV	d	N
$1i_{13/2}$	-10.22	-10.25	1.14	5
$1h_{9/2}$	-12.60	-12.66	1.1	4
$1h_{11/2}$	-15.61	-15.59	0.8	20
$1f_{5/2}$	-26.52	-26.58	0.92	5
$1f_{7/2}$	-28.20	-28.28	0.88	5
$1s_{1/2}$	-41.63	-41.58	0.6	5

can be calculated including only $N = 5$ iterations (with the exception for state $1h_{11/2}$) and the following terms in expansion (11.23) tend to zero. We got the correct eigenvalues for the quantum number $n = 1$ only after $N = 3$ iterations. In this case, oscillations of eigenfunctions U_α at large distances are excluded. From the calculations presented in Table 11.2, we obtain fast convergence of eigenfunctions and the obvious eigenfunctions because without convergence of (11.17), convergence of (11.21) is not possible.

11.4   Conclusions

Comparing the energy levels or eigenvalues E_{nLj} presented in Table 11.1 with $\bar{E}_{nLj}$ calculated by the discretization method using the standard program EIGEN A[14], we obtained good coincidences. In this chapter, the potential representation method, earlier used for positive A[4] and negative energies using physical and nonphysical nonperturbed solutions, has been applied for the calculation of negative energies or eigenvalues for the first time using only physical solutions (wave functions). It is very important for the comparison of nuclei scattering experiments A[4] and bound states in the shell models. Applying the variation of parameter d of the model potential for minimizing the number N of iterations, we got sufficiently exact eigenvalues E_{nLj} after the first iteration. We obtained the correct value of the principal quantum number $n = 1$ by avoiding oscillations of eigenfunctions at large distances only after $N = 3$ iterations, whereas, after the first and the second iterations, we obtained principal quantum numbers $n = 2$. However, we obtained the incorrect n value only as a result of oscillations of wave functions at large distances, which did not have any influence on eigenvalues. Practical calculations verified by Theorem 1 showed that integral equations (11.17) are correct and can be solved for principal quantum number $n = 1$ only.

References

1. Byron Jr, F. W. and Fuller, R. W. (1992). *Mathematical of Classical Physics*, Dover Publications, New York.
2. Flügge, S. (1999). *Practical Quantum Mechanics I*, Springer-Verlag, Berlin.
3. Guter, R. S. and Janpolski, A. R. (1976). *Differential Equations*, Vyssheya Shkola, Moscow.
4. Oganessian, Y. (2010). Heaviest nuclei, *Nucl. Phys. A*, **834**(1), 331–337.
5. Ortega, J. M. and Pool, V. G. (1981). *An Introduction to Numerical Methods for Differential Equation*, Pitman Publishing Inc., New York.
6. Pahlavani, R. M. and Alavi, S. A. (2012). Solutions of Woods–Saxon potential with spin–orbit and centrifugal terms, *Theor. Phys.* **58**(5), 739–743.
7. Shirokov, Yu. M. and Judin, N. P. (1982). *Nuclear Physics*, Vol. 1, Mir, Moscow, p. 447.

Chapter 12

The Perturbation Theory
for Bound States

12.1 Introduction

The Schrödinger equation can be solved analytically only for the one- and two-particle case for some potentials. The most important analytical solutions were found for harmonic oscillator and Coulomb potentials. For more complicated potentials and atoms beginning from He, the bound-state perturbation theory must be used. In this chapter, we will investigate the Rayleigh–Schrödinger perturbation theory [1] with the aim to compare this classical theory with the potential representation method. This theory represents the solutions for the time-independent Schrödinger equation as follows, which are defined by the energy eigenvalues E_n corresponding to degenerated or unique eigenvectors ψ_n:

$$\hat{H}_p \Psi_{pn} = E_{pn} \Psi_{pn}, \quad n = 1, 2, 3, \ldots \tag{12.1}$$

when the perturbed Hamiltonian consists of two parts

$$\hat{H}_p = \hat{H}_0 + gV = H_0 + V_1 \cdot 0 \le g \le 1. \tag{12.2}$$

We present the unperturbed equation when $g = 0$ in following way:

$$\hat{H}_0 \Psi_{0n} = E_{0n} \Psi_{0n}. \tag{12.3}$$

For small g, we have small changes of unperturbed eigenvalues

$$E_{pn} = E_{0n} + \Delta E_{0n}, \tag{12.4}$$

and eigenvectors

$$\Psi_{pn} = \Psi_{0n} + \Delta\Psi_{0n}. \tag{12.5}$$

ΔE_{0n} can be evaluated after substituting (12.4) and (12.5) into (12.1). Considering the equal largeness terms, we obtain for the first-order approach,

$$\Delta E_{0n} \approx \int \Psi_{0n}^* V_1 \Psi_{0n} dx^3. \tag{12.6}$$

Solutions of perturbed equations can be expanded by the complete orthonormal set of eigenfunctions

$$\Psi_{pn} = \Psi_{0n} + g\Psi_{1n} = g^2\Psi_{2n} + \cdots \tag{12.7}$$

and the corresponding eigenvalues

$$E_{pn} = E_{0n} + gE_{1n} + g^2 E_{2n} + \cdots. \tag{12.8}$$

Substituting the two expansions (12.7) and (12.8) into (12.1) and comparing the terms at equal power, $n = 0, 1, 2$ of g^n, we obtain

$$H_0\Psi_{0n} = E_{0n}\Psi_{0n}, \tag{12.9}$$

$$H_0\Psi_{1n} + V\Psi_{0n} = E_{0n}\Psi_{1n} + E_{1n}\Psi_{0n}, \tag{12.10}$$

$$H_0\Psi_{2n} + V\Psi_{1n} = E_{0n}\Psi_{2n} + E_{1n}\Psi_{1n} + E_{2n}\Psi_{0n}, \tag{12.11}$$

The obtained Eq. (12.9) coincides with (12.3). Now we can rewrite (12.10) as follows:

$$(H_0 - E_{0n})\Psi_{1n} = (E_{1n} - V)\Psi_{0n}. \tag{12.12}$$

Expanding Ψ_{1n} by the sum of the unperturbed wave functions Ψ_{0m},

$$\Psi_{1n} = \sum_{m=0}^{\infty} c_{1nm}\Psi_{0m}, \tag{12.13}$$

and substituting in (12.12), we obtain

$$\sum_{m=0}^{\infty} c_{1nm}(E_{0m} - E_{0n})\Psi_{0m} = (E_{1n} - V)\Psi_{0n}. \tag{12.14}$$

Multiplying (12.14) on Ψ_{0m}^* and taking the integral over all space τ, we obtain

$$c_{1nm}(E_{0m} - E_{0n}) \cdot \delta_{0m,0n} = E_{1n}\delta_{0m,0n} - \int \Psi_{0m}^* V \Psi_{0n} d\tau,$$

$$(12.15)$$

when we used (12.11) for expanding the set of unperturbed orthonormal functions satisfying the following condition:

$$\int \Psi_{0m}^* \Psi_{0n} d\tau = \delta_{0m,0n}, \, \delta_{0m,0n} = 1(0m = 0n), 0(0m \neq 0n). \quad (12.16)$$

We used the Kronecker delta equal to 1 when $0m = 0n$ and 0 when $0m \neq 0n$. For the first case from (12.15), we obtain the first-order correction of the average value of perturbation

$$E_{1n} = V_{mn} = \int \Psi_{0m}^* V \Psi_{0n} d\tau, \qquad (12.17)$$

and for the second case,

$$c_{1nm} = \frac{V_{mn}}{E_n - E_m}, \quad m \neq n, \, 0m = m, \, 0n = n. \qquad (12.18)$$

Using (12.18) and (12.17), we obtain the first-order perturbed wave function

$$\Psi_{pn} = \Psi_{0n} + \Psi_{1n} = \sum_{m=0}^{\infty} c_{1nm} \Psi_{0m}, \quad n \neq m \qquad (12.19)$$

and energy

$$E_{pn} = E_{0n} + E_{1n} = E_{0n} + V_{nn}, \quad V_{nn} = g \int \Psi_{0n}^* V \Psi_{0n} d\tau \quad (12.20)$$

in the first approach. We must remark that coefficient c_{1nn} cannot be defined using the formula (12.19). The coefficients c_{1nn} can be found from the orthonormality of the obtained perturbed wave functions as follows:

$$\int \Psi_{pn}^* \Psi_{pm} d\tau = \delta_{nm}. \qquad (12.21)$$

As a matter of fact, the smallest of the quantity of the values of the perturbation energies V_{nn} and convergence velocity of series (12.19) are defined by the condition

$$|V_{nn}| \ll |E_{0n}|, \quad \frac{|V_{mn}|}{|E_n - E_m|} \ll 1. \tag{12.22}$$

The second approach of perturbation for the wave function can be obtained from (12.11) by introducing

$$\Psi_{2n} = \sum_{m=0}^{\infty} c_{2nm} \Psi_{0m}, \tag{12.23}$$

and requiring orthonormality for the wave function (12.23) using the definition (12.21). Also, the expression for energy improvements [2] can be obtained as follows:

$$E_{2n} = \sum_{m=0}^{\infty} \frac{|V_{nm}|^2}{E_n - E_m}, \quad n \neq m. \tag{12.24}$$

Then, the perturbed energy after the introduction of (12.11) and (12.24) will be given as

$$E_{pn} = E_{0n} + E_{1n} + E_{2n}. \tag{12.25}$$

For calculating V_{nm} in (12.24), we must use the wave function (12.19). The degenerated case is that where one eigenvalue E_{0n} of the unperturbed Schrödinger's equation has $n \neq m$ number of eigenfunctions. In this case, the unique corresponding eigenfunction for free solution can be obtained by a linear combination of degenerated functions

$$\Psi_{0n} = \sum_{m=1}^{j} C_{0nm} \Psi_{0m}. \tag{12.26}$$

Applying Eq. (12.12) and the presented expansion, we can obtain solutions for the degenerated case in a more complicated form [2].

12.2   Standard Green's Functions

The main idea of the standard Green's functions is applied for solving the differential equation

$$\hat{L}R = f \tag{12.27}$$

with a linear differential operator and given function f. The $\hat{L}$ possesses an orthonormal set of eigenfunctions $\phi_n(r)$. Using the expression of the differential equation when $f = 0$, we have

$$\hat{L}\phi_n(r) = \lambda_n\phi_n(r). \tag{12.28}$$

With a complete orthonormal set of unperturbed functions $\phi_n(r)$, we can obtain the infinite number of elements which represent the following expansions:

$$R(r) = \sum_{n=1}^{\infty} a_n\phi_n(r), \tag{12.29}$$

$$f(r) = \sum_{n=1}^{\infty} \beta_n\phi_n(r). \tag{12.30}$$

If we substitute (12.29) and (12.30) into Eq. (12.28), we can obtain [3]

$$\sum_{n=1}^{\infty} (a_n\lambda_n - \beta_n)\phi_n = 0. \tag{12.31}$$

Since $\phi_n(r)$ is linearly independent, [3] we conclude from (12.31) that

$$a_n = \beta_n/\lambda_n, \quad R(r) = \sum_{n=1}^{\infty} \frac{1}{\lambda_n}\beta_n\phi_n(r), \quad \beta_n = (\phi_n, f). \tag{12.32}$$

Now we can write the solution in more detail as follows:

$$R(r) = \sum_{n=1}^{\infty} \frac{1}{\lambda_n}\phi_n(r) \int \phi_n^*(r_1)f(r_1)dr_1 = \int \sum_{n=1}^{\infty} \frac{\phi_n(r)\phi_n^*(r_1)}{\lambda_n} f(r_1)dr_1. \tag{12.33}$$

The last formula can be written in the following way:

$$R(r) = \int G(r, r_1) f(r_1) dr_1, \tag{12.34}$$

where the standard Green's functions can be expressed in terms of symmetric infinite expansion, i.e.,

$$G(r, r_1) = \sum_{n=1}^{\infty} \frac{\phi_n(r) \phi_n^*(r_1)}{\lambda_n}, \quad G(r, r_1) = G(r_1, r)^*. \tag{12.35}$$

Usually, perturbations withdraw the degeneration of wave functions $\phi_n(r)$. The solutions of the perturbed equation (12.27) in our case can be found from the integral equation (12.34) and eigenvalues of (12.27) calculations by the iteration method.

References

1. Merzcbacher, E. (1970). *Quantum Mechanic*, John Wiley & Sons, New York, p. 621.
2. Schiff, L. I. (1955). *Quantum Mechanics*, McGraw-Hill Book Company, Inc. New York, Toronto, London, p. 473.
3. Byron Jr, F. W. and Robert, W. Fuller (1992). *Mathematics of Classical and Quantum Physics*, Dover Publications, Inc., New York, p. 661.

Chapter 13

The Perturbation Method of Variation of Free Constants

13.1 Green's and Undefined Functions

We will consider a general solution for the following differential equation:

$$y'' + p_1(x)y' + p_2(x)y = q(x), \quad Ly = q(x), \qquad (13.1)$$

with linear ordinary differential operator L. Instead of using the free constants for the expression of general solution of Eq. (13.1) when coefficients p_1, p_2 do not depend on x, we can use [1] undefined functions C_1, C_2 and linearly independent solutions y_1, y_2 of (13.1) for $q(x) = 0$, i.e.,

$$y = C_1(x)y_1(x) + C_2(x)y_2(x). \qquad (13.2)$$

Introducing the additional condition [1]

$$C_1'y_1 + C_2'y_2 = 0, \qquad (13.3)$$

we obtain the first derivative

$$y' = C_1y_1' + C_2y_2', \qquad (13.4)$$

and the second derivative

$$y'' = C_1y_1'' + C_2y_2'' + C_1'y_1' + C_2'y_2', \qquad (13.5)$$

of the general solution y. Substituting (13.2), (13.4) and (13.5) into the differential equation (13.1) and keeping in mind the fact that y_1, y_2 satisfy

$$y'' = p_1(x)y' + p_2(x)y = 0, \tag{13.6}$$

we obtained

$$C_1'y_1' + C_2'y_2' = q(x). \tag{13.7}$$

Solving (13.3) and (13.7), we get the expression for the derivatives of indefinite coefficients

$$C_1' = \frac{1}{W}q(x)y_2, \quad C_2' = -\frac{1}{W}q(x)y_1 \tag{13.8}$$

by the Wronskian

$$W = y_1'y_2 - y_1y_2', \quad W \neq 0, \tag{13.9}$$

of linearly independent solutions y_1, y_2 of (13.6). Integrating (13.8), we obtain the expressions of coefficients

$$C_1 = \int \frac{1}{W}q(x)y_2dx + K_1, \quad C_2 = -\int \frac{1}{W}q(x)y_1dx + K_2, \tag{13.10}$$

which define the general solution (13.2) by coefficients (13.10) C_1, C_2

$$y = y_1 \int \frac{1}{W}q(x)y_2dx - y_2 \int \frac{1}{W}q(x)y_1dx + K_1y_1 + Ky_2. \tag{13.11}$$

The obtained equation transforms the differential equation in the integral equation, which can be solved by the iteration method by finding solutions or eigenfunctions and eigenvalues of the Schrödinger equation for negative energies. This method is more exact than the discretization method [2] or the standard method of Green's functions [3] expressed by the infinite series.

Let suppose that the operator L possesses a complete orthonormal set of eigenfunctions $\varphi_n(x)$, i.e.,

$$L\phi_n(x) = \lambda_n \phi_n(x), \tag{13.12}$$

with eigenvalues λ_n. Then we can write

$$y(x) = \sum_{n=1}^{\infty} \alpha_n \phi_n(x), \tag{13.13}$$

$$q(x) = \sum_{n=1}^{\infty} \beta_n \phi_n(x). \tag{13.14}$$

Substituting expansions (13.13) and (13.14) into (13.1), we obtain

$$\sum_{n=1}^{\infty} (\alpha_n \lambda_n - \beta_n) \phi_n = 0. \tag{13.15}$$

When ϕ_n are linearly independent functions, we obtain the equation

$$\alpha_n = \beta_n / \lambda_n, \tag{13.16}$$

and expression for the solution

$$y(x) = \sum_{n=1}^{\infty} \frac{1}{\lambda_n} \phi_n(x) \beta_n$$

$$= \sum_{n=1}^{\infty} \frac{1}{\lambda_n} \phi_n(\phi_n, q) = \sum_{n=1}^{\infty} \frac{1}{\lambda_n} \phi_n(x) \int \phi_n^*(x') q(x') dx'. \tag{13.17}$$

In this case, the last formula can be obtained in the following way:

$$y(x) = \int \sum_{n=1}^{\infty} \frac{\phi_n(x) \phi_n^*(x')}{\lambda_n} q(x') dx' = \int G(x, x') q(x') dx' \tag{13.18}$$

where the following Green's function was introduced:

$$G(x, x') = \sum_{n=1}^{\infty} \frac{\phi_n(x) \phi_n^*(x')}{\lambda_n}, \tag{13.19}$$

like in the infinity series. In this case, the solution (13.17) and expansion (13.18) are operating with the infinity series and a convergence problem for these expansions is complicated.

References

1. Guter, R. S. and Janpolski, A. R. (1976). *Differential Equations*, Vyssheja Shkola, Moscow, p. 304 (in Russian).
2. Ortega J. M. and Pool, V. G. (1981). *An Introduction to Numerical Methods for Differential Equations*, New York, Pitman Publishing Inc., p. 288.
3. Frederick, W. Byron, Jr. and Fuller, R. W. (1992). *Mathematics of Classical and Quantum Physics*, Dover Publications, Inc., New York, p. 661.

Chapter 14

Green's Functions and Non-physical Solutions

14.1 Introduction

We discuss the multiplicative perturbation theory for the discrete energies. In this case, we can solve the Schrödinger equation by parts. The multiplicative perturbation theory for the positive energies was developed using the modified Lagranges' method and was published in our previous papers. Green's functions have been expressed by unperturbed wave functions and non-physical solutions A[9] for the harmonic oscillator and Coulomb potentials. Usually, Green's functions are expressed by an infinite sum where a complete, orthonormal set of unperturbed eigenfunctions is used, but in this case, we obtained the expression containing only one term. It is important that problems do not occur when unperturbed solutions are degenerated. We obtain the non-degenerated eigenfunctions in the first-step iteration procedure, because the kernel of the obtained integral equation is directly proportional to the perturbation potential. This method is also used for calculating the tails of wave functions and eigenvalues with high accuracy.

This chapter develops the modified method of Lagrange [1] and the multiplicative perturbation theory A[5, 6] for positive energies.

It is well known that the solution of some linear non-homogeneous differential equation can be expressed by Green's function [2].

Usually, Green's function is represented as an infinite expansion in terms of the eigenfunctions of the homogeneous equation.

But a more handy method for obtaining Greens' functions is the bound-state perturbation theory proposed in Ref. A[9]. We can construct Green's function from regular $\lim_{r\to 0} u_0 r^{-L-1} = 1$ and irregular $\lim_{r\to 0} f_0 r^L = 1$ solutions of the radial unperturbed Schrödinger equation. In this chapter, the irregular solutions for the radial harmonic oscillator potential

$$V_1(r) = \frac{m\omega^2 r^2}{2}, \tag{14.1}$$

and the Coulomb potential as well as the energies of the bound states for the Woods–Saxon potential are obtained. The harmonic oscillator potential was used as a model potential added to and subtracted from the Hamiltonian. Green's functions are expressed by the regular and irregular degenerate solutions of the Schrödinger equation for the harmonic oscillator potential. If only physical solutions for the model or unperturbed potential are unknown, then Green's functions may be obtained in a standard form of an infinite sum, where a complete set of unperturbed eigenfunctions must be used. In this case, the solutions for the perturbed potential may appear to be not sufficiently exact.

14.2   Non-physical Solutions of the Radial Schrödinger Equation

Here, we consider the non-physical solutions for the harmonic oscillator and Coulomb potential. First, we consider the solutions of the Schrödinger equation

$$\frac{d^2}{dr^2} U_\alpha + \frac{L(L+1)}{r^2} U_\alpha + C(E_\alpha - V(r))U_\alpha = 0, \tag{14.2}$$

where $C = \frac{2m}{\hbar^2}$ for the harmonic oscillator potential (14.1). The wave function is presented as follows [3]:

$$U_{nL} = e^{-\frac{\rho}{2}} \rho^{\frac{L+1}{2}} \sum_{k=0}^{n-1} \alpha_k \rho^k, \tag{14.3}$$

where $\rho = \frac{m\omega r^2}{\hbar^2}$, $n = 1, 2, 3, \ldots$, and

$$a_{k+1} = \frac{k - n + 1}{(k+1)\left(k + L + \frac{3}{2}\right)} a_k. \tag{14.4}$$

The value $a_0 = 1$ is chosen in the recurrence relation (14.4). Using similar methods but different boundary conditions, we obtain the other linearly independent solution for Eq. (14.3), i.e.,

$$F_{nL} = e^{-\frac{\rho}{2}} \rho^{-\frac{L}{2}} w(\rho), \quad w = \sum_{k=0}^{\infty} b_k \rho^k, \tag{14.5}$$

$$b_{k+1} = \frac{k - \frac{1}{2}\left(\varepsilon_{nL} + L - \frac{1}{2}\right)}{(k+1)\left(k - L + \frac{1}{2}\right)} b_k, \quad b_0 = 1. \tag{14.6}$$

The eigenvalues are also known as

$$E_{nL} = \varepsilon_{nL} \hbar \omega, \quad \varepsilon_{nL} = 2n + L - \frac{1}{2}. \tag{14.7}$$

The asymptotic behavior of the wave function $U_\alpha(r)$ and non-physical linearly independent solution $F_\alpha(r)$ at the origin are given as follows:

$$\lim_{r \to 0} U_{nL} = \left(r^2 \frac{m\omega}{\hbar}\right)^{\frac{L+1}{2}}, \quad \lim_{r \to 0} F_{nL} = \left(r^2 \frac{m\omega}{\hbar}\right)^{-\frac{L}{2}}. \tag{14.8}$$

Using these asymptotic expressions, we can obtain the Wronskian of the linearly independent solutions U_{nL} and F_{nL} as follows:

$$W_0 = (2L + 1)\left(\frac{m\omega}{\hbar}\right)^{\frac{1}{2}}. \tag{14.9}$$

Now let us find the nonphysical solutions of the Schrödinger equation (14.2) for the Coulomb attraction potential, i.e.,

$$V(r) = \frac{A}{r}, \quad A = -\frac{Ze^2}{4\pi\varepsilon_0}. \tag{14.10}$$

The physical solution in this case can be expressed in the following standard form [4]:

$$U_{c,nL} = e^{-k_n r} r^{L+1} \sum_{v=0}^{N} b_v r^v, \tag{14.11}$$

where $k_n = -\frac{CA}{2n}$, $n = N + L + 1$, and

$$b_{v+1} = \frac{2k_v(v + L + 1) + CA}{(v + 1)(v + 2L + 2)} b_v, \quad b_0 = 1. \tag{14.12}$$

Let us suppose that $b_{N+1} = 0$, then

$$k_n = -\frac{CA}{2n}, \quad n = N + L + 1, \tag{14.13}$$

where n is the principal quantum number.

The nonphysical solutions can be expanded into a negative power series. So, we obtain

$$F_{c,nL} = e^{k_n r} r^{L+1} \sum_{v=0}^{2L+1} C_{-v} r^{-v}, \tag{14.14}$$

$$C_{-v-1} = \frac{v(v - 2L - 1)}{2k_v(v - 2L - 1) + CA} C_{-v}. \tag{14.15}$$

In Ref. [5], it was shown that the non-physical (irregular) solution could be obtained after substituting $-n$ and $-l - 1$ instead of n and l in the physical solution. This result was important for the subsequent investigations of non-physical solutions. In addition, we see that

$$\lim_{r \to 0} U_{c,nL} r^{-L-1} = 1, \quad \lim_{r \to 0} F_{c,nL} r^L = C_{-2L-1}. \tag{14.16}$$

For derivation of the integral equation, we must obtain the Wronskian, and from Eq. (14.16), we obtain

$$W_{0,c} = (2L + 1)C_{-L-1}. \tag{14.17}$$

A more complicated algebraic expression of the non-physical solution has been presented in Ref. [6].

14.3   Derivation of the Integral Equation

If we consider the Woods–Saxon potential

$$V_s(r) = -V_0(1 + \exp\{\alpha(r - R)\})^{-1} \tag{14.18}$$

and spin–orbit potential

$$V_{sl}(r) = -\kappa \frac{1}{r} \frac{d}{dr} V_s(r)(\sigma l) \tag{14.19}$$

as the perturbations and use the harmonic oscillator potential (14.1) as a model potential, we must solve Eq. (14.2) for the potential

$$V(r) = V_D(r) + V_1(r), \quad V_D(r) = V_s(r) + V_{sl}(r) - V_1(r). \quad (14.20)$$

The eigenfunctions of Eq. (14.2) in the case of the multiplicative perturbation theory [2, 3] are expressed by multiplying the eigenfunction U_{nL} for the model potential $V_1(r)$ with the factor function $\varphi_{2,nLj}$ determined by the potential $V_D(r)$

$$U_\alpha = \varphi_{2,nLj} U_{nL}. \quad (14.21)$$

Substituting Eq. (14.21) into Eq. (14.2), we obtain the radial Schrödinger equation in the potential A[3, 6] representation

$$U_{nL} \frac{d^2}{dr^2} \varphi_2 + 2 \left(\frac{d}{dr} \varphi_2 \right) \left(\frac{d}{dr} U_{nL} \right) - CV_\delta U_{nL} \varphi_2 = 0, \quad (14.22)$$

where

$$V_\delta(r) = V_D(r) - \Delta E_{nLj}, \quad (14.23)$$

$$E_\alpha = E_{nL} - \Delta E_{nLj}. \quad (14.24)$$

Using the modified Lagrange method, equation of the same type (Ref. A[5]) has been reduced to the integral equation

$$\varphi_2 U_{nl} = U_{nL}\beta + \frac{F_{nL}}{W_0} \int_0^r U_{nL} CV_\delta \varphi_2 U_{nL} dx$$

$$- \frac{U_{nL}}{W_0} \int_r^\infty F_{nL} CV_\delta \varphi_2 U_{nL} dx, \quad (14.25)$$

$$\beta = 1 + \frac{1}{W_0} \int_0^\infty F_{nL} CV_\delta \varphi_2 U_{nL} dx.$$

The obtained equation may be written in the following way:

$$\varphi_2(r) U_{nL}(r) = U_{nL}(r)\beta + \int_0^\infty G(r > r_1, r < r_1) CV_\delta \varphi_2 U_{nL} dr_1, \quad (14.26)$$

where, according to Ref. [2], the kernel of the integral equation is Greens' function

$$G(r > r_1, r < r_1) = -\frac{1}{W_0} F_{nL}(r > r_1) U_{nL}(r < r_1). \quad (14.27)$$

For the bound states, the solutions $\varphi_2 U_{nL}$ are regular and decreases to zero at infinity. Using the boundary condition from Eq. (14.25), we get

$$\Delta E_{nLj} = \frac{\int_0^\infty U_{nL} V_D \varphi_{2,nLj} U_{nL} dr}{\int_0^\infty U_{nL} \varphi_{2,nLj} U_{nL} dr}. \tag{14.28}$$

Equation (14.26) can be reduced to a more convenient form as follows:

$$\varphi_2 U_{nL} = U_{nL}\beta + \frac{U_{nL}}{W_0} \int_0^r F_{nL} C V_\delta \varphi_2 U_{nL} dr_1$$

$$- \frac{F_{nL}}{W_0} \int_0^r U_{nL} C V_\delta \varphi_2 U_{nL} dr_1. \tag{14.29}$$

The integral equations (14.28) and (14.29) can be solved by the iteration method. For the zero approximation on the right-hand side of Eqs. (14.28) and (14.29) we must take $\varphi_2 = 1$, and then find ΔE_{nLj} from Eq. (14.28) and $\varphi_2 U_{nL}$ from Eq. (14.29). The model potential may be chosen freely, but it is better, if the unperturbed wave functions are close to the perturbed wave functions $\varphi_2 U_{nL}$. Then the small number of the iterations ensures the results of high accuracy. The frequency ω_0 for the model harmonic oscillator potential can be determined by the root-mean-square radius of the nuclei [7] as follows:

$$\omega_0 = 41 A^{-\frac{1}{3}} \frac{\text{MeV}}{\hbar}, \quad \omega = d\omega_0. \tag{14.30}$$

The constant d has been found by demanding the minimum of the energy. The test calculations are in good agreement with the results obtained in this chapter where the discretization method [8] has been used.

14.4   Results and Conclusions

The integral equation (14.29) has been solved for the Woods–Saxon potential (14.18) and the spin–orbit potential (14.19). The one-nucleon energy levels of ^{197}Au for neutrons with the parameters $V_0 = 43.8\,\text{MeV}$, $\alpha = 1.4286\,\text{fm}^{-1}$, $R = 1.26 A^{\frac{1}{3}}\,\text{fm}$, $\kappa = 0.353\,\text{fm}^2$

have been found. These results have been compared with the energies obtained by the discretization method [8], $B = 14\,\text{fm}$,

$$\frac{U_{i+1} - 2U_i + U_{i-1}}{h^2} + \rho(r_i)U_i = \lambda U_i, \quad \lambda = -bE,$$

$$i = 0, 1, 2, \ldots, n, h = \frac{B}{n}, \quad (14.31)$$

with the boundary conditions $U(0) = U(B) = 0$. We have obtained the eigenfunctions and energies by solving the system of these equations with the help of the program EIGEN [9]. For both methods, we used a step $h = 0.2$ and the upper limit of the interval region is $B = 14\,\text{fm}$. The energies $(E_{nLj})^d$ obtained by the discretization method coincide with the energies E_{nLj} obtained by solving the integer equation (14.29) with sufficient accuracy. The results are presented in Table 14.1. The potential representation method A[15] is very useful for calculating the tails of the wave functions with a high accuracy.

The probability density of the obtained wave functions at infinity is about 10^{-6} of their maximum values and without oscillations. This

Table 14.1. The energy levels $(E_{nLj})^d$ of neutrons obtained by the discretization method using the program EIGEN [9] and E_{nLJ} calculated by solving the integral equations (14.29) and (14.28) after N iterations.

nLj	E_{nLj} Mev	E_{nLJ} MeV	d	N
$1s_{\frac{1}{2}}$	-39.03	-39.36	1.00	6
$2s_{\frac{1}{2}}$	-28.05	-28.42	1.00	14
$3s_{\frac{1}{2}}$	-13.42	-13.98	1.00	16
$1p_{\frac{1}{2}}$	-34.90	-34.80	0.90	6
$2p_{\frac{1}{2}}$	-21.30	-21.23	0.90	6
$3p_{\frac{1}{2}}$	-5.963	-5.997	0.90	9
$1p_{\frac{3}{2}}$	-35.31	-35.05	1.00	6
$2p_{\frac{3}{2}}$	-22.08	-22.07	1.00	5
$3p_{\frac{3}{2}}$	-6.871	-6.830	1.00	5
$1d_{\frac{3}{2}}$	-29.54	-29.50	0.85	8
$2d_{\frac{3}{2}}$	-13.96	-13.95	0.85	17

fact is very important in the calculations of the relativistic corrections for the mass in the average field of neutron and proton shells A[15]. Within zero approximation, the analytical expression of the wave functions has well-defined derivatives of the fourth order, and the relativistic mass corrections can be defined with the same accuracy. Then the same approaches were calculated using Eq. (14.31). When the Schrödinger equation is solved by the discretization method [8], it is difficult to calculate the wave functions and their fourth-order continuous derivatives with sufficient accuracy. The perturbation method proposed in this chapter is very useful for calculating the relativistic corrections, which depend strongly on the tails of the wave functions for the nuclei and atoms.

The breakdown of the simple Rayleigh–Schrödinger theory in the case of degenerate unperturbed states, when the perturbation expansion becomes meaningless, is the main problem for the spectra of atoms [10] and scattering matrices calculations [11]. At the same time, the perturbation method presented can be applied without modification, if we use the non-physical solutions for the Coulomb potential for the calculations in atomic spectroscopy.

References

1. Zeldovich, J. B. (1985). Perturbation theory for one dimension task in quantum mechanics and method lagrange, *JETP*, **31**, 1101 (in Russian).
2. Byron Jr, F. W. and Fuller, R. W. (1992). *Mathematics of Classical and Quantum Physics*, Dover Publications, Inc., New York, p. 661.
3. Solovjov, V. G. (1981). *Theory of Atomic Nucleus, Nuclear Models*, Energoizdat, Moscow, p. 295.
4. Merzcbacher, E. (1970). *Quantum Mechanics*, John Wiley & Sons, New York, p. 621.
5. Savukynas, A. J., Karosienė A. V., *et al.* (1965). *Symmetry of Mirror Reflection for Radial Integrals Case*, Mokslas, Vilnius.
6. Veselov, M. G. and Labzovski, L. N. (1986). *Theory of Atom, Structure of Electronic Shells*, Nauka, Moscow, p. 327.
7. Jelley, N. A. (1990). *Fundamentals of Nuclear Physics*, Cambridge University Press, New York, p. 278.
8. Ortega, J. M. and Pool, V. G. (1981). *An Introduction to Numerical Methods for Differential Equations*, New York, Pitman Publishing Inc., p. 288.
9. *System/360 Scientific Subroutine Package, Programmer's Manual*, IBM Technical, p. 224.

10. Merkelis, G. V., Kaniauskas, J. M. and Rudzikas, Z. B. (1985). Formal methods of stationary theory of perturbation in atoms, *Lithuanian Phys. Coll.*, **25**(5), 21.

11. Au, C. K. and Chow, Ch.-K. (1997). Perturbation of scattering phase shifts in one dimensions: Closed form results," *Pilnas Pavadinimas Phys. Lett. A*, **226**, 327.

Chapter 15

The Potential Representation Method for Non-spherical Perturbations

15.1 Introduction

This method of potential representation is presented for non-spherical perturbation potential A[8]. The system of integral equations for finding the eigenfunctions and eigenvalues is obtained. These equations can be applied to nuclear physics where no integer values of orbital momentum for definition of quantum states can be used. This method can be used for calculations of eigenfunctions and eigenvalues for nucleons of non-spherical nuclei.

Here, the new perturbation method is applied for considering of the changes in binding energies of deformed nuclei in the one-particle shell model. Formerly, we discussed A[4] the multiplicative perturbation theory for the one-particle discrete energies in the spherical symmetrical case. However, frequently particles move in an anisotropic potential well.

The Hamiltonian in the non-spherical case can be written in the following manner:

$$H = H_0 + V(r, \theta, \varphi), \tag{15.1}$$

where the spherical symmetric part of the Hamiltonian is

$$H_0 = -\frac{\hbar^2 \Delta_r}{2\mu} - \frac{\hbar^2 \Delta_{\theta,\varphi}}{2\mu r^2} + V_0(r). \tag{15.2}$$

For the non-spherical potential $V(r, \theta, \varphi)$, variables cannot be separated, but for the wave function

$$\Psi_\alpha = \Psi_\alpha(r, \theta, \varphi), \tag{15.3}$$

we can introduce diagonal terms of the density matrix

$$R_\alpha^*(r)R_\alpha(r) = \int_0^{2\pi} d\varphi \int_0^\pi \Psi_\alpha^*(r, \theta, \varphi)\Psi_\alpha(r, \theta, \varphi)\sin\theta d\theta, \tag{15.4}$$

$$\Theta_\alpha^*(\theta)\Theta_\alpha(\theta) = \int_0^{2\pi} d\varphi \int_0^\infty \Psi_\alpha^*(r, \theta, \varphi)\Psi_\alpha(r, \theta, \varphi)r^2 dr, \tag{15.5}$$

$$\Phi_\alpha^*(\varphi)\Phi_\alpha(\varphi) = \int_0^\pi \sin\Theta d\Theta \int_0^\infty \Psi_\alpha^*(r, \theta, \varphi)\Psi_\alpha(r, \theta, \varphi)r^2 dr. \tag{15.6}$$

R_α evidently depends on the principal quantum number for a spherical potential and on the averaged parameters or the suitable quantum numbers which depend on the angular coordinates. Θ_α and Φ_α depend on the averaged parameters r, ϕ and θ, r, respectively.

The approximate wave function can be represented in the following way:

$$\Psi_{\alpha_1, \alpha_2, \alpha_3} = R_{\alpha_1}(r)\Theta_{\alpha_2}(\theta)\Phi_{\alpha_3}(\varphi). \tag{15.7}$$

In the case of $V(r, \theta, \phi) = 0$, we have

$$R_{0, \alpha_1} = R_{nl}(r), \tag{15.8}$$

$$\Theta_{0, \alpha_2} = \Theta_{l, m_l}(\theta), \tag{15.9}$$

$$\Phi_{0, \alpha_3} = \Phi_m(\varphi). \tag{15.10}$$

The Legendre polynomials $\Theta_{l, m_l}(\theta)$ and the function Φ_m can be found in any study of quantum mechanics.

15.2 Integral Equations for Negative Energies in the Potential Representation

We can write the following integral [3]:

$$I = \int_0^\infty \int_0^\pi \int_0^{2\pi} R_{\alpha_1}^* \Theta_{\alpha_2}^* \Phi_{\alpha_3}^* H R_{\alpha_1} \Theta_{\alpha_2} \Psi_{\alpha_3} r^2 dr \sin\theta d\theta d\varphi$$

$$H = -\frac{\hbar^2 \Delta_r}{2\mu} + V_0(r) + V(r, \theta, \varphi). \tag{15.11}$$

For every energy E_α, the non-degenerate eigenfunctions must be orthogonal and can be normalized to unity as follows:

$$\int_0^\infty R_{\alpha_1}^*(r)R_{\alpha_{1,1}}(r)r^2dr = \delta_{\alpha_1,\alpha_{1,1}}, \tag{15.12}$$

$$\int_0^\pi \Theta_{\alpha_2}^*(\theta)\Theta_{\alpha_{2,1}}(\theta)r^2\sin\theta d\theta = \delta_{\alpha_2,\alpha_{2,1}}, \tag{15.13}$$

$$\int_0^{2\pi} \Phi_{\alpha_3}^*(\varphi)\Phi_{\alpha_{3,1}}(\varphi)r^2\sin\varphi d\varphi = \delta_{\alpha_3,\alpha_{3,1}}. \tag{15.14}$$

Solving the variation problem for the integral (15.11) and varying the functions R_{α_1}, Θ_{α_2}, Φ_{α_3} in the integrals (15.12)–(15.14) with Lagrangian multipliers $-E_{\alpha_1}$, $-E_{\alpha_2}$, $-E_{\alpha_3}$ we obtain the following system of equations:

$$\frac{d^2}{dr^2}u_{\alpha_1} + (-(l_{\alpha_2,\alpha_3})^2 - cV_0(r) - cV_{\alpha_2,\alpha_3}(r) + cE_{\alpha_1})u_{\alpha_1} = 0,$$

$$c = \frac{2\mu}{\hbar^2}, \tag{15.15}$$

$$-\frac{\hbar^2}{2\mu}a_{\alpha_1}\left(\frac{1}{\sin\theta}\left(\frac{d}{d\theta}\sin\theta\frac{d}{d\theta}\Theta_{\alpha_2}\right) - \frac{m_{\alpha_3}^2}{\sin^2\theta}\Theta_{\alpha_2}\right)$$
$$+ V(\theta)_{\alpha_1,\alpha_3}\Theta_{\alpha_2} - E_{\alpha_2}\Theta_{\alpha_2} = 0, \tag{15.16}$$

$$-\frac{\hbar^2}{2\mu}a_{\alpha_1}b_{\alpha_2}\frac{d^2}{d\varphi^2}\Phi_{\alpha_3} + V_{\alpha_1,\alpha_2}(\varphi)\Phi_{\alpha_3} - E_{\alpha_3}\Phi_{\alpha_3} = 0. \tag{15.17}$$

Here, the following designations are used:

$$R_{\alpha_1}(r) = \frac{u_{\alpha_1}(r)}{r}, \tag{15.18}$$

$$l_{\alpha_2,\alpha_3}^2 = -\int_0^\pi\int_0^{2\pi}\Theta_{\alpha_2}^*\Phi_{\alpha_3}^*\Delta_{\theta,\varphi}\Theta_{\alpha_2}\Phi_{\alpha_3}\sin\theta d\theta d\varphi, \tag{15.19}$$

$$a_{\alpha_1} = \int_0^\infty R_{\alpha_1}^*\frac{1}{r^2}R_{\alpha_1}r^2dr, \tag{15.20}$$

$$m_{\alpha_3}^2 = -\int_0^{2\pi}\Phi_{\alpha_3}^*\frac{d^2}{d\varphi^2}\Phi_{\alpha_3}d\varphi, \tag{15.21}$$

$$b_{\alpha_2} = \int_0^\pi\Theta_{\alpha_2}^*\frac{1}{\sin^2\theta}\Theta_{\alpha_2}\sin\theta d\theta, \tag{15.22}$$

$$V_{\alpha_2,\alpha_3}(r) = \int_0^\pi \int_0^{2\pi} \Theta_{\alpha_2}^* \Phi_{\alpha_3}^* V(r,\theta,\varphi) \Theta_{\alpha_2} \Phi_{\alpha_3} \sin\theta d d\varphi, \quad (15.23)$$

$$V_{\alpha_1,\alpha_3}(\theta) = \int_0^\infty \int_0^{2\pi} R_{\alpha_1}^* \Phi_{\alpha_3}^* V(r,\theta,\varphi) R_{\alpha_1} \Phi_{\alpha_3} r^2 dr d\varphi, \quad (15.24)$$

$$V_{\alpha_1,\alpha_2}(\varphi) = \int_0^\infty \int_0^\pi R_{\alpha_1}^* \Theta_{\alpha_2}^* V(r,\theta,\varphi) R_{\alpha_1} \Theta_{\alpha_2} r^2 dr \sin\theta d\theta. \quad (15.25)$$

Equations (15.16) and (15.17) can be reduced to the standard form. Introducing a new function

$$P_{\alpha_2}(x) = (1-x^2)\Theta_{\alpha_2}(x), \quad x = \cos\vartheta, \quad (15.26)$$

we can present Eqs. (15.16) and (15.17) in the following form:

$$\frac{d^2}{dx^2} P_{\alpha_2} + \left(\frac{1}{1-x^2} + \frac{x^2 - m_{\alpha_3}^2}{(1-x^2)^2} \right.$$

$$\left. - \frac{c_{\alpha_1}}{1-x^2} V_{\alpha_1,\alpha_3}(\theta) + \frac{c_{\alpha_1}}{1-x^2} E_{\alpha_2} \right) P_{\alpha_2} = 0, \quad (15.27)$$

where

$$c_{\alpha_1} = \frac{2\mu}{\hbar^2 a_{\alpha_1}}, \quad (15.28)$$

and

$$\frac{d^2}{d\varphi^2} \Phi_{\alpha_3} + c_{\alpha_1,\alpha_2}(-V_{\alpha_1,\alpha_2}(\varphi) + E_{\alpha_3}) \Phi_{\alpha_1} = 0, \quad (15.29)$$

where

$$c_{\alpha_1,\alpha_2} = \frac{2\mu}{\hbar^2 a_{\alpha_1} b_{\alpha_2}}. \quad (15.30)$$

Now, substituting

$$E_{\alpha_1} = E_{0,\alpha_1} + \Delta E_{\alpha_1}, \quad (15.31)$$

$$c_{\alpha_1} E_{\alpha_2} = c_{0,\alpha_1} E_{0,\alpha_2} + c_{\alpha_1} \Delta E_{\alpha_2}, \quad (15.32)$$

$$c_{\alpha_1} E_{\alpha_2} = c_{0,\alpha_1} E_{0,\alpha_2} + c_{\alpha_1} \Delta E_{\alpha_2}, \quad (15.33)$$

$$L_{\alpha_2,\alpha_3}^2 = l(l+1) + \Delta(l_{\alpha_2,\alpha_3})^2, \quad (15.34)$$

$$m_{\alpha_3}^2 = m^2 + (\Delta m_{\alpha_3})^2, \quad (15.35)$$

into Eqs. (15.15), (15.27) and (15.29), we obtain

$$\frac{d^2}{dr^2} u_{\alpha_1} - \frac{l(l+1)}{r^2} u_{\alpha_1}$$

$$+ c(E_{0,\alpha_1} - V_0(r) - V_{1,\alpha_2,\alpha_3}(r)) u_{\alpha_1} = 0, \qquad (15.36)$$

$$\frac{d^2}{dx^2} P_{\alpha_2} + \left(\frac{c_{0,\alpha_1}}{1-x^2} E_{0,\alpha_2} + \frac{1}{1-x^2} + \frac{x^2 - m^2}{(1-x^2)^2} \right.$$

$$\left. - \frac{c_{\alpha_1}}{1-x^2} V_{\alpha_1,\alpha_3}(\theta) \right) P_{\alpha_2} = 0, \qquad (15.37)$$

$$\frac{d^2}{dr^2} \Phi_{\alpha_3} + (c_{0,\alpha_1,\alpha_2} E_{0,\alpha_3} - c_{\alpha_1,\alpha_2} V_{1,\alpha_2,\alpha_3}(\varphi)) \Phi_{\alpha_3} = 0, \quad (15.38)$$

where

$$V_{1,\alpha_2,\alpha_3} = V_{\alpha_2,\alpha_3}(r) - \Delta E_{\alpha_1} + \frac{(\Delta l_{\alpha_2,\alpha_3})^2}{cr^2}, \qquad (15.39)$$

$$V_{1,\alpha_2,\alpha_3}(\theta) = V_{\alpha_1,\alpha_3}(\theta) - \Delta E_{\alpha_2} + \frac{(\Delta m_{\alpha_3})^2}{c_{\alpha_1}(1-x^2)}, \qquad (15.40)$$

$$V_{1,\alpha_1,\alpha_2}(\varphi) = V_{\alpha_1,\alpha_2}(\varphi) - \Delta E_{\alpha_3}. \qquad (15.41)$$

E_{0,α_1} is the eigenvalue of the energy for the spherical potential $V_0(r)$ for the state 0, α_1 for $V_{\alpha_2,\alpha_3}(\theta) = 0$. In the case of $V_{\alpha_2,\alpha_3}(\theta) = 0$, $V_{\alpha_1,\alpha_2}(\varphi) = 0$ the eigenvalues of the eigenfunctions (15.9), (15.10) are known

$$E_{0,\alpha_2} = a_{0,\alpha} \frac{l(l+1)\hbar^2}{2\mu}, \qquad (15.42)$$

$$E_{0,\alpha_3} = a_{0,\alpha_1} b_{0,\alpha_2} \frac{m^2 \hbar^2}{2\mu}. \qquad (15.43)$$

The system of Eqs. (15.36)–(15.38) can be solved using the method described in the paper A[9]. The solutions can be presented in the following way:

$$u_{\alpha_1}(r) = c_{1,\alpha_1}(r) f_{0,\alpha_1}(r)$$

$$+ c_{2,\alpha_2}(r) u_{0,\alpha_2}(r), \quad u_{\alpha_1}(r) = \varphi_{2,\alpha_1}(r) u_{0,\alpha_1}(r), \quad (15.44)$$

$$P_{\alpha_2}(r) = c_{1,\alpha_2}(x)q_{lm}(x)$$
$$+ c_{2,\alpha_2}(x)P_{lm}(x), \quad P_{\alpha_2}(x) = \varphi_{2,\alpha_2}(r)P_{lm}(x), \quad (15.45)$$
$$\Phi_{\alpha_3}(\varphi) = c_{1,\alpha_3}(\varphi)\Phi_m^*(\varphi)$$
$$+ c_{2,\alpha_3}(\varphi)\Phi_m(\varphi), \quad \Phi_{\alpha_3}(\varphi) = \varphi_{2,\alpha_3}(\varphi)\Phi_m(\varphi), \quad (15.46)$$

$f_{0,\alpha_1}(r)$, $u_{0,\alpha_1}(r)$ are unperturbed linearly independent solutions of Eq. (15.36) with boundary conditions $\lim_{r \to 0} f_{0,\alpha_1}(r)r' = 1$, $\lim_{r \to 0} u_{0,\alpha_1}(r)r^{-l-1} = 1$. P_{lm} and q_{lm} are expressed by the associated Legendre polynomials of the first and second kind in the following way [3]:

$$P_{l,m_l} = (1 - x^2)^{1/2}\Theta_{l,m_l}, \quad (15.47)$$
$$q_{l,m_l} = (1 - x^2)^{1/2}Q_{l,m_l}, \quad (15.48)$$

which are linear independent solutions of Eq. (15.37) in the case $V_{1,\alpha_1,\alpha_3}(\theta) = 0$. The linearly independent solutions of Eq. (15.38) for $V_{1,\alpha_1,\alpha_2}(\theta) = 0$ are

$$\Phi_{m_l}^*(\varphi) = \frac{1}{\sqrt{2\pi}}e^{-im_l\varphi}, \quad (15.49)$$

$$\Phi_{m_l}(\varphi) = \frac{1}{\sqrt{2\pi}}e^{im_l\varphi}. \quad (15.50)$$

Demanding that the following boundary conditions — corresponding to Green's functions — would be satisfied:

$$\lim_{r \to 0} u_{\alpha_1}(r)r^{-l-1} = 1, \quad (15.51)$$

$$\lim_{x \to 1} P_{\alpha_2}(x) = P_{lm}(x), \quad (15.52)$$

$$\lim_{\varphi \to 0} \Phi_{\alpha_3}(\varphi) = \Phi_m(\varphi) \quad (15.53)$$

and performing the procedure presented in the paper A[9], we obtain integral equations with the corresponding Greens' functions

$$\varphi_{2\alpha_1}(r)u_{0,\alpha_1}(r) = u_{0,\alpha_1}(r)\beta_{\alpha_1} + \int_0^\infty G(r > r_1, r < r_1)$$

$$\times cV_{1,\alpha_2,\alpha_3}\varphi_{2,\alpha_1}u_{0,\alpha_1}dr_1, \quad (15.54)$$

$$\beta_{\alpha_1} = 1 + \frac{1}{f_0(\alpha_1)} \int_0^\infty f_{0,\alpha_1} V_{1,\alpha_2,\alpha_3} \varphi_{2,\alpha_1} u_{0,\alpha_1} dr_1, \quad (15.55)$$

$$G(r > r_1, r < r_1) = -\frac{f_{0,\alpha_1}(r > r_1) u_{0,\alpha_1}(r < r_1)}{f_0(\alpha_1)}, \quad (15.56)$$

$$\varphi_{2\alpha_2}(x) P_{lm}(x) = P_{lm}(r)\beta_{\alpha_2} + \int_1^{-1} G(x > x_1, x < x_1)$$
$$\times \frac{c_{\alpha_1}}{1 - x_1^2} V_{1,\alpha_1,\alpha_3} \varphi_{2,\alpha_2} P_{lm} dx_1, \quad (15.57)$$

$$\beta_{\alpha_2} = 1 + \frac{1}{f_0(\alpha_2)} \int_0^\infty q_{lm} \frac{c_{\alpha_1}}{1 - x_1^2} V_{1,\alpha_1,\alpha_3} \varphi_{2,\alpha_2} P_{lm} dx_1, \quad (15.58)$$

$$G(x > x_1, x < x_1) = -\frac{q_{lm}(x > x_1) P_{lm}(x < x_1)}{f_0(\alpha_2)}. \quad (15.59)$$

$$\varphi_{2\alpha_3}(\varphi) \Phi_m(\varphi) = \Phi_m(\varphi)\beta_{\alpha_3} + \int_0^{2\pi} G(\varphi > \varphi_1, \varphi < \varphi_1)$$
$$\times c_{\alpha_1,\alpha_2} V_{1,\alpha_1,\alpha_2} \varphi_{2,\alpha_3} \Phi_m d\varphi_1, \quad (15.60)$$

$$\beta_{\alpha_3} = 1 + \frac{1}{f_0(\alpha_3)} \int_0^{2\pi} \Phi_m^* c_{\alpha_1,\alpha_2} V_{1,\alpha_1,\alpha_2} \varphi_{2,\alpha_3} \Phi_m d\varphi_1, \quad (15.61)$$

$$G(\varphi > \varphi_1, \varphi < \varphi_1) = -\frac{\Phi_m^*(\varphi > \varphi_1) \Phi_m(\varphi < \varphi_1)}{f_0(\alpha_3)}. \quad (15.62)$$

Here, $f_0(\alpha_1)$, $f_0(\alpha_2)$, $f_0(\alpha_3)$ are the appropriate Wronskians.

For bound states, the solutions of Eqs. (15.54), (15.57) and (15.60) must satisfy the following boundary conditions A[9]:

$$\lim_{r \to \infty} \varphi_{2,\alpha_1}(r) u_{0,\alpha_1}(r) = 0, \quad \lim_{x \to -1} \varphi_{2,\alpha_1}(x) P_{lm}(x) = \text{const}, \quad (15.63)$$

and the conditions of periodicity

$$\varphi_{2\alpha_2}(\theta) P_{lm}(\theta) = \varphi_{2,\alpha_2}(\theta + 2\pi) P_{lm}(\theta + 2\pi), \quad (15.64)$$

$$\varphi_{2\alpha_3}(\varphi) \Phi_m(\varphi) = \varphi_{2,\alpha_3}(\varphi + 2\pi) \Phi_m(\varphi + 2\pi). \quad (15.65)$$

From here, we obtain

$$\Delta E_{\alpha_1} = \frac{\int_0^\infty u_{0,\alpha_1} \left(V_{\alpha_2,\alpha_3}(r) + \frac{(\Delta l_{\alpha_2,\alpha_3})^2}{cr^2} \right) \varphi_{2,\alpha_1} u_{0,\alpha_1} dr}{\int_0^\infty u_{0,\alpha_1} \varphi_{2,\alpha_1} u_{0,\alpha_1} dr}, \quad (15.66)$$

$$\Delta E_{\alpha_2} = \frac{\int_1^{-1} \Theta_{lm} \left(V_{\alpha_1,\alpha_3}(\theta) + \frac{(\Delta m_{\alpha_3})^2}{c_{\alpha_1}(1-x^2)} \right) \varphi_{2,\alpha_2} \Theta_{lm} dx}{\int_1^{-1} \Theta_{lm} \varphi_{2,\alpha_2} \Phi_{lm} dx}, \quad (15.67)$$

$$\Delta E_{\alpha_3} = \frac{\int_0^{2\pi} \Phi_m^* V_{\alpha_1,\alpha_2} \varphi_{2,\alpha_3} \Phi_m d\varphi}{\int_0^{2\pi} \Phi_m^* \varphi_{2,\alpha_3} \Phi_m d\varphi}. \quad (15.68)$$

Equation (15.66) is obtained taking into account Eq. (15.47). Its eigenfunctions and eigenvalues can be found using Eqs. (15.54), (15.57), (15.60) and (15.65)–(15.67), i.e.,

$$W(P_l^m(x), Q_l^m(x)) = \frac{e^{im\pi} 2^{2m} \Gamma\left(\frac{l+m+2}{2}\right) \Gamma\left(\frac{l+m+1}{2}\right)}{(1-x^2)\Gamma\left(\frac{l-m+2}{2}\right) \Gamma\left(\frac{l-m+1}{2}\right)}, \quad (15.69)$$

and also using Eqs. (15.45) and (15.48), we obtain the Wronskian $f(\alpha_2)$

$$W(P_{lm}, Q_{lm}(x)) = \frac{2^{2m} \Gamma\left(\frac{l+m+2}{2}\right) \Gamma\left(\frac{l+m+1}{2}\right)}{\Gamma\left(\frac{l-m+2}{2}\right) \Gamma\left(\frac{l-m+1}{2}\right)} \left(\frac{2(l+1)(l-m)}{2(l+m)} \right)^{\frac{1}{2}}.$$
$$(15.70)$$

We can find the Wronskian $f(\alpha_3)$

$$W(\Phi_m^*, \Phi_m) = \frac{im}{\pi}, \quad (15.71)$$

from Eqs. (15.49) and (15.50). Using the correlation [2]

$$Q_l^{-m} = e^{-2im\pi} \frac{\Gamma(l-m+1)}{\Gamma(l+m+1)} Q_l^m, \quad (15.72)$$

we can find the Wronskian $f(\alpha_2)$ for negative m, i.e.,

$$W(P_{l,-m}, Q_{l,-m}) = e^{-2im\pi} \frac{\Gamma(l-m+1)}{\Gamma(l+m+1)} W(P_{lm}, Q_{lm}). \quad (15.73)$$

The integral equations (15.54), (15.57), (15.60) and Eqs. (15.66)–(15.68) were obtained A[9] for finding the eigenfunctions and the corrections ΔE_{α_1}, ΔE_{α_2} and ΔE_{α_3} to unperturbed eigenvalues E_{0,α_1}, E_{0,α_2} and E_{0,α_3}. Using Eqs. (15.19) and (15.21), the eigenvalues of l_{α_2,α_3}^2 and $m_{\alpha_3}^2$ must be found in the process of solving Eqs. (15.54), (15.57) and (15.60). For the spherically symmetrical potential, they are equal to $l(l+1)$ and m^2, respectively. The integral equations and three-dimensional Green's functions, (15.56), (15.59) and (15.62), can be modified for scattering and many-particle cases. This method does not require the use of infinite series. For the first approach, we must take $\varphi_{2,\alpha_1} = 1$, $\varphi_{2,\alpha_3} = 1$, $(\Delta l_{\alpha_2,\alpha_3}) = 0$, $\Delta m_{\alpha_3} = 0$.

By comparing Eq. (15.15) with Eqs. (15.1) and (15.2), we see that the energy of the particle in non-symmetrical potential $V(r, \theta, \varphi)$, coincides with E_{α_1}. The Wronskian $f_0(\alpha_1)$ depends on the unperturbed potential $V_0(r)$. The obtained equations can be used for calculations of one-particle state energies for non-spherical nuclei [3].

References

1. Shiff, L. L. (1995). *Quantum Mechanics*, New York, Toronto, London, McGraw-Hill, p. 473.
2. Abramowitz, M. (1964). *Handbook of Mathematical Functions*, ed. J. A. Stegun, New York, National Bureau of Standards, p. 832.
3. Solovjow, V. G. (1981). *Theory of Atomic Nuclei, Nuclear Models*, Moscow, Energoizdat, p. 296 (in Russian).

Chapter 16

Solutions with the Model Potential for the Potential Representation Method

16.1 Introduction

A general solution of the Schrödinger equation in the potential representation has been obtained in the form of integral equations. In this representation, the wave function for positive and negative energies or bound states can be expressed as a product of the unperturbed solution for model potential and the function which depends on the additional potential or potential perturbation. Here, we have proved that this method A[9] is equivalent to the method of variation of constants for negative energies. The linearly independent solutions of the Schrödinger equation for harmonic oscillator potential have been obtained for derivation of integral equations, which are used for finding eigenfunctions and eigenvalues for the Woods–Saxon potential. Eigenvalues, obtained by numerical iterations, of these integral equations are in good agreement with the results obtained by the discretization method. The kernels of the obtained integral equations are proportional to the perturbation or difference of the Woods–Saxon and harmonic oscillator potentials.

We have obtained the analytical solutions of the one-particle radial Schrödinger equation for negative energies in the form of the integral equations by using the potential representation method proposed for positive A[2, 5, 6] and negative A[15] energies. The main idea of this method expresses the radial wave function as a product of the model solution and the function, which depends on the difference of interaction potentials. This chapter is a generalization of integral equations derived in Ref. A[1] for positive energies, where the radial wave functions can be obtained by multiplying the free solution on function which depends on the perturbation potential. In Ref. A[1], the following postulate was presented: after adding the potential to the Hamiltonian, a new radial wave function can be obtained by multiplying the unperturbed or model wave function with the function that depends on the added potential. We have proved this postulate A[6, 9, 15] and presented it like a theorem using the method of indefinite coefficients. From this theorem, the connection between the scattering matrices for short-range potentials and for Coulomb potentials has been obtained A[6]. In potential representation A[2, 7, 9, 15], the perturbed radial function $U_{\alpha nL}(r)$ with the set of quantum numbers αnL can be expressed as a product of the unperturbed wave function $U_{0nL}(r)$ with the set of quantum numbers nL, i.e.,

$$U_{\alpha nL}(r) = \varphi_\alpha(r)U_{0nL}(r), \qquad (16.1)$$

on the multiplier $\varphi_\alpha(r)$ depending on an additional potential. The main problem considered in this chapter is the interpretation and convergence of the eigenvalues obtained using the harmonic oscillator potential for modelling of the Woods–Saxon potential. At large distances, these potentials have essentially different asymptotical behavior, i.e., r^2 and $\exp\{-\alpha r\}$, $\alpha > 0$, respectively.

16.2   Modelling the Solutions of the Schrödinger Equation with the Harmonic Oscillator Potential

We consider the Schrödinger equation

$$\frac{d^2}{dr^2}U_{\alpha nL} - \frac{L(L+1)}{r^2}U_{\alpha nL} + c(E_{0nL} + \Delta E_\alpha - V_D(r) - V_0(r))U_{\alpha nL}$$
$$= 0, \qquad (16.2)$$

where $c = \frac{2\mu}{\hbar^2}$, and the potentials are defined as

$$V_0(r) = \frac{m\omega^2 r^2}{2} - V_S(0), \quad V_D(r) = V_S(r) - V_0(r),$$

$$V_S(r) = -V(1 + \exp\{\alpha(r - R)\})^{-1}. \tag{16.3}$$

Here, we have added and subtracted the harmonic oscillator potential $V_0(r)$. The energies are expressed by the sum of E_{0nL} for a model potential $V_0(r)$ and energy changes ΔE_α for perturbation $V_D(r)$. We are interested in finding eigenvalues $E_{0nL} + \Delta E_\alpha$ and eigenfunctions $U_{\alpha nL}$ of Eq. (2.1) with perturbed potential $V_D(r)$ by using unperturbed analytical physical U_{0nL} and linearly independent non-physical F_{0nL} solutions and eigenvalues E_{0nL} of the Schrödinger equation for the model potential $V_0(r)$. For the harmonic oscillator potential, we have obtained the following solutions [3]:

$$U_{0nL} = \exp\left\{-\frac{\rho}{2}\right\} \rho^{\frac{L+1}{2}} \sum_{k=0}^{n-1} a_k \rho^k, \quad \rho = \frac{m\omega r^2}{h}, \quad n = 1, 2, 3, \tag{16.4}$$

$$a_{k+1} = \frac{k - \frac{1}{2}\left(\varepsilon_{nL} - L - \frac{3}{2}\right) m\omega r^2}{(k+1)\left(k + L + \frac{3}{2}\right)} a_k, \quad a_0 = 1, \tag{16.5}$$

$$F_{0nL} = \exp\left\{-\frac{\rho}{2}\right\} \rho^{-\frac{L}{2}} w(\rho), \quad w(\rho) = \sum_{k=0}^{\infty} b_k \rho^k, \tag{16.6}$$

$$b_{k+1} = \frac{k - \frac{1}{2}\left(\varepsilon_{nL} + L - \frac{1}{2}\right) m\omega r^2}{(k+1)\left(k - L + \frac{1}{2}\right)} b_k, \quad b_0 = 1, \tag{16.7}$$

and eigenvalues

$$E_{0nL} = \varepsilon_{nL} h\omega - V_S(0), \quad \varepsilon_{nL} = 2n + L - \frac{1}{2}, \tag{16.8}$$

$$\omega - d\omega_0, \quad \omega_0 = 41 A^{-\frac{1}{3}} \frac{MeV}{h}.$$

Frequencies ω_0 for the harmonic oscillator potential can be determined A[5] from the radius R of nucleus and depend on a number of nucleons A and $V_S(0) = -V_0(0)$ from (16.3). The coefficients of power series $w(\rho)$ for very high powers satisfy the abbreviated recursion relation $b_{k+1}/b_k \to \frac{1}{k}$, $k \gg 1$.

From this ratio and [4], A[9] and (16.4), we can make a conclusion that for $\rho \to \infty$, we have

$$w(\rho) \to \exp\{\rho\}, \quad F_{0nL} \to \exp\left\{\frac{1}{2}\rho\right\}. \tag{16.9}$$

Using the asymptotical expressions $\lim_{\rho\to 0} U_{0nL} = \rho^{L+1}$ and $\lim_{\rho\to 0} F_{0nL} = \rho^{-L}$, we can obtain the Wronskian A[15]

$$W_0(F_{0nL}, U_{0nL})$$

$$= F_{0nL}\frac{dU_{0nL}}{dr} - U_{0nL}\frac{dF_{0nL}}{dr}, \quad W_0 = (2L+1)\left(\frac{m\omega}{h}\right)^{\frac{1}{2}}. \tag{16.10}$$

The perturbed $U_{\alpha nL}$ and unperturbed U_{0nL} solutions must have the same boundary condition at the origin r^{L+1} and at the infinity

$$\lim_{r\to 0} \varphi_\alpha = 1, \ \lim_{r\to\infty} \varphi_\alpha(r)U_{0nL}(r) = 0. \tag{16.11}$$

16.3   Green's Functions for the Potential Representation

The multiplicative or potential representation perturbation theory can be realized by using the modified method of Lagrange A[9], [2, 7]. In this case, the perturbed solution $U_{\alpha nL}$ of Eq. (16.1) for perturbation potential $V_D(r)$ can be expressed in the following way:

$$U_{\alpha nL} = C_1(r)F_{0nL}(r) + C_2(r)U_{0nL}(r), \tag{16.12}$$

where F_{0nL} and U_{0nL} are linearly independent solutions for model potential $V_0(r)$. Using the same factorization as in Ref. A[20], i.e.,

$$U_{\alpha nL} = \left(C_1(r)\frac{F_{0nL}(r)}{U_{0nL}(r)} + C_2(r)\right)U_{0nL}(r),$$

$$U_{\alpha nL}(r) = \varphi_\alpha(r)U_{0nL}(r), \tag{16.13}$$

and the additional condition [7] for derivatives of indefinite coefficients

$$C_1'(r)\frac{F_{0nL}(r)}{U_{0nL}(r)} + C_2'(r) = 0, \tag{16.14}$$

we obtain the derivatives

$$\varphi'_\alpha = -W_0 \frac{C_1(r)}{U^2_{0nL}(r)}, \quad W_0 = F_{0nL} U'_{0nL} - U_{0nL} F'_{0nL}, \quad (16.15)$$

$$\varphi''_\alpha = -W_0 \left(\frac{C'_1(r)}{U^2_{0nL}(r)} - 2C_1(r) \frac{U'_{0nL}(r)}{U^3_{0nL}(r)} \right). \quad (16.16)$$

Here, W_0 is the Wronskian of linearly independent solutions $F_{0nL}(r)$ and U_{0nL}. Substituting (16.13), (16.15) and (16.16) into (16.2) and including the expression (16.14), we obtain

$$C'_{1\alpha} = -\frac{1}{W_0} U_{0nL} cV_d \varphi_\alpha U_{0nL}, \quad V_d = V_D - \Delta E_\alpha, \quad (16.17)$$

$$C'_{2\alpha}(r) = -\frac{1}{W_0} F_{0nL} cV_d \varphi_\alpha U_{0nL}. \quad (16.18)$$

Integrating the last equation and taking into account the boundary conditions (16.11), we obtain

$$C_{1\alpha}(r) = -\frac{1}{W_0} \int_0^r U_{0nL} cV_d \varphi_\alpha U_{0nL} dr_1,$$

$$C_{2\alpha}(r) = \beta_\alpha - \frac{1}{W_0} \int_r^\infty F_{0nL} cV_d \varphi_\alpha U_{0nL} dr_1, \quad (16.19)$$

$$\beta_\alpha = 1 + \frac{1}{W_0} \int_0^\infty F_{0nL} cV_d \varphi_\alpha U_{0nL} dr_1.$$

Substituting these equations into (16.12), we obtain the integral equation for perturbed eigenfunctions as follows:

$$\varphi_\alpha U_{0nL} = U_{0nL} \beta_\alpha - \frac{F_{0nL}}{W_0} \int_0^r U_{0nL} cV_d \varphi_\alpha U_{0nL} dr_1$$

$$- \frac{U_{0nL}}{W_0} \int_r^\infty F_{0nL} U_{0nL} cV_d \varphi_\alpha U_{0nL} dr_1. \quad (16.20)$$

The presented equation can be rewritten as

$$\varphi_\alpha U_{0nL} = U_{0nL} \beta_\alpha + \int_0^\infty G_0(r_\rangle, r_\langle) V_d \varphi_\alpha U_{0nL} dr_1,$$

where according to A[9] the kernel of this integral equation is Green's function

$$G_0(r_\rangle, r_\langle) = -\frac{1}{W_0} F_{0nL}(r_\rangle) U_{0nL}(r_\langle). \tag{16.21}$$

Using (16.19), we can reduce Eq. (16.20) to a more convenient form

$$\varphi_\alpha U_{0nL} = U_{0nL} + \frac{U_{0nL}}{W_0} \int_0^r F_{0nL} c V_d \varphi_\alpha U_{0nL} dr_1$$

$$- \frac{F_{0nL}}{W_0} \int_0^r U_{0nL} c V_d \varphi_\alpha U_{0nL} dr_1, \tag{16.22}$$

$$V_d(r) = V_S(r) - V_0(r) - \Delta E_\alpha.$$

Here, we have the kernels of integral equation

$$M_1(r) = F_{0nL}(r) V_d(r), \quad M_2(r) = U_{0nL}(r) V_d(r). \tag{16.23}$$

Requiring that the perturbed solution $\varphi_\alpha U_{0nL}$ would satisfy the boundary condition (16.11) and taking into account Eq. (16.9), we obtain the following expression for the calculation of perturbed eigenvalues at N-th iteration:

$$E_{\alpha nLN} = E_{0nL} + \Delta E_{\alpha N}, \Delta E_{\alpha N} = \frac{\int_0^\infty U_{0nL} V_D(r) \varphi_{\alpha N} U_{0nL} dr}{\int_0^\infty U_{0nL} \varphi_{\alpha N} U_{nL} dr},$$

$$V_D(r) = V_S(r) - V_0(r). \tag{16.24}$$

The integral equations (16.22), (16.24) can be solved by the iteration method. For the first iteration of $\Delta E_{\alpha nL0}$ and $\varphi_{\alpha 0} U_{0nL}$ in the integrals (16.24) and the right-hand side of (16.22), we must take $\varphi_{\alpha 0} = 1$, $N = 0$. For calculation of more exact values of energy $E_{\alpha nL}$ and perturbed eigenfunctions, we must solve the integral equations using the obtained values of $\Delta E_{\alpha 0}$ and $\varphi_{\alpha 1} U_{0nL}$.

16.4　The Accuracy of the Solutions Obtained

Theorem 1. *The changes ΔE_α of the unperturbed eigenvalues E_{0nL} and the potential representation functions φ_α depend on the perturbation of potential energy V_d and weakly depend on the parameter of the integral equation.*

Proof. Convergence of eigenvalues $E_{\alpha n L}$ depends on the absolute values of the kernels (16.23) of the integral equation (16.22). For that consideration, we can present the integral equation (16.22) as follows:

$$\varphi_\alpha U_{0nL} = U_{0nL} + \lambda \left(U_{0nL} \int_0^z F_{0nL} V_d \varphi_\alpha U_{0nL} dz_1 \right.$$

$$\left. - F_{0nL} \int_0^z U_{0nL} c V_d \varphi_\alpha U_{0nL} dz_1 \right), \quad \lambda = \frac{cF^2}{W_0}, \quad (16.25)$$

where we have introduced a new dimensionless variable z expressed by the F-Fermi unit 10^{-15} m. From (16.8) and (16.10), we calculate the values of the parameters $\lambda_1 = 0{,}73701 \cdot 10^{-3}/\sqrt{d}$ and $\lambda_2 = 0{,}79991 \cdot 10^{-3}/\sqrt{d}$ of the integral equation (16.25) for the neutrons of nucleus $A_1 = 208$ and $A_2 = 340$. Taking into account that $d \in [0{,}6; 1{,}2]$, we can use the method proposed in Ref. [6], where the solution of the integral equation can be expressed by power series of λ. Using this method, we suppose that for a sufficiently large number of iterations N, we can present approximately

$$\varphi_\alpha = \sum_{m=0}^N \varphi_{m\alpha} \lambda^m, \quad \varphi_{0\alpha} = 1. \quad (16.26)$$

For the first approach, we take $\varphi_{0\alpha} = 1$ in (16.24), and for the energy improvement $\Delta E_{\alpha 0}$ in (16.24) not depending on λ, we obtain $1 + \varphi_{1\alpha}\lambda$. Then we obtain that $\varphi_{1\alpha}$ does not depend on λ, i.e.,

$$\lambda \varphi_{1\alpha} U_{0nL} = \lambda \left(U_{0nL} \int_0^z F_{0nL} V_{d0} U_{0nL} dz_1 \right.$$

$$\left. - F_{0nL} \int_0^z U_{0nL} c V_{d0} U_{0nL} dz_1 \right), \quad (16.27)$$

but depends only on differences $V_{dN}(r) = V_S(r) - V_0(R) - \Delta E_{\alpha,N}$, $V_0(r) = m\omega^2 r^2/2 - V_S(0)$ of the Woods–Saxon (16.3) $V_S(r)$ and model $V_0(r)$ potentials for a variable parameter d *in* $\omega = d\omega_0$. The U_{0nL} and F_{0nL} are physical and non-physical solutions A[15] of the Schrödinger equation for the model potential. Substituting (16.26) into (16.25) and (16.24) for any iterations N, we obtain

$$\varphi_{N+1,\alpha} U_{0nL} = U_{0nL} \int_0^r F_{0nL} c V_{dN} \varphi_{N\alpha} U_{0nL} dz$$

$$- F_{0nL} \int_0^r U_{0nL} c V_{dN} \varphi_{N\alpha} U_{0nL} dz,$$

$$V_{dN}(r) = V_S(r) - V_0(r) - \Delta E_{\alpha,N}, \qquad (16.28)$$

$$E_{\alpha,N} = \int_0^\infty U_{0nL} V_D U_{0nL} dr + \lambda \int_0^\infty U_{0nL} V_D \varphi_{1\alpha} U_{0nL} dr$$

$$+ \lambda^N \int_0^\infty U_{0nL} V_D \varphi_{N\alpha} U_{0nL} dr.$$

It is well known in quantum mechanics that the first iteration for ΔE_α of perturbation calculations is the most important for a small perturbation [3]. Providing variation of d or ω, we can precisely solve the integral equations (16.24) and (16.25) for the small potential perturbation $V_D = V_S - V_0$ in the region $z \in [0; R_m]$ with the minimum number of iterations N. The second iteration for ΔE_α will be obtained substituting in (16.24) approximate function $\varphi_\alpha = 1 + \lambda \varphi_{1\alpha}$ depending on λ (16.25), (16.8) or $\sqrt{d}$. We have the kernels (16.23) of integral equation (16.25) where perturbations of the energies ΔE_α depending on the quantum states are included. In this case, the optimal values of parameter d must be different for different quantum states or energy levels $E_{\alpha nL}$.

In Ref. [6], it was proved that the solutions of the integral equations like (16.25) converge uniformly, when the following conditions are satisfied:

$$|U_{0nL}| < N_U, U_{0nL} \int_0^{R_m} F_{0nL} V_{dN} dZ - F_{0nL} \int_0^{R_m} U_{0nL} V_{dN} dZ < M.$$
$$(16.29)$$

Thus, in our case, we should prove that the following conditions:

$$|\varphi_{m\alpha} U_{0nL}| < N_U M^m, \quad |\lambda| < \frac{1}{M} - \varepsilon, \quad (\varepsilon > 0) \qquad (16.30)$$

are satisfied. We can normalize A[16] the eigenfunctions U_{0nL} and then we have in (16.29) that $N_U < 1$. Theoretical evaluation of M is a hard task, but experimentally, we have obtained the convergence of eigenvalues (16.24), (16.28) $\Delta E_{\alpha nLN}$ and eigenfunctions (16.25) $\varphi_\alpha U_{0nL}$ for $\frac{0{,}73701 \cdot 10^{-3}}{\sqrt{d}} \leq \lambda \leq \frac{0{,}79991 \cdot 10^{-3}}{\sqrt{d}}$ when $d \in [0{,}6; 1{,}2]$. For small values of ω or parameter d, the eigenvectors $E_{\alpha nL}$ depend on perturbation V_D and very weakly on λ, parameter M has a finite value. Theorem 1 is proved.

It follows that the model potential $V_0(r)$ for $r \to R_m$ is increasing like r^2 but U_{0nL} are rapidly decreasing like $\exp\{-Cr^2\}$. We have a convergence of eigenvalues $E_{\alpha nL}$ and eigenfunctions for Eqs. (16.24) and (16.25) at large distances R_m when Theorem 1 is satisfied and eigenvalues weakly depend on parameter λ.

In computational experiments, we have used different limits of integration R_m in (16.24) and (16.25) from 0 till 1.5R or 2.4R, where $R = 1{,}24 A^{\frac{1}{3}}$ is a nucleus radius expressed in Fermi units. Integration limits are increasing for decreasing energies of nucleons $|E_{\alpha nL}|$ A[15]. In order to get approximate solutions, we must find the eigenvalues which only weakly depend on d in some region of Δd, because the exact eigenvalues do not depend on the parameters ω of the model potential. Practically, we can find approximate solutions by changing $\omega = d\omega_0$ in the interval $d \in [0{,}6; 1{,}2]$ A[15]. We must choose solutions giving minimum values of $|\Delta E_{\alpha nL}|$ and minimum iteration numbers for the calculation of eigenvalues with some given accuracy.

In Ref. A[9], calculations of energies E_{nLj} were done for the Woods–Saxon potential of one-nucleon energy levels for neutrons of nucleus ^{197}Au. Integral equations were solved with the variation of parameters d and ω of the model potential. The energies E_{nLj} expressed in MeV were compared with the eigenvalues $\bar{E}_{nLj}$ obtained by the discretization method, in the interval $14\,\mathrm{fm}(10^{-15}\,\mathrm{m})$ with the step $0{.}2\,\mathrm{fm}$ by using the EIGEN program [5]. Results for quantum states with principal quantum numbers n, orbital–momentum quantum numbers $L = 0$ (s state) and resulting momentum quantum numbers $j = 1/2$ are presented in Table 16.1. Here N is the number of iterations.

The same energy levels were calculated using the EIGEN program A[14] and by solving integral equations (16.24) and (16.25), with the parameter d and ω of the model potential (16.3) for the nucleus ^{208}Pb. The results are presented in Table 16.2.

Table 16.1.   The energy levels $\bar{E}_{nLj}$ and E_{nLj} of neutrons of nucleus ^{197}Au.

nLj	$\bar{E}_{nLj}$, MeV	E_{nLj}, MeV	d	N
$1s_{\frac{1}{2}}$	-39.03	-39.36	1.00	6
$2s_{\frac{1}{2}}$	-28.05	-28.42	1.00	14
$3s_{\frac{1}{2}}$	-13.42	-13.98	1.00	18

Table 16.2.   The energy levels $\bar{E}_{nLj}$ and E_{nLj} of neutrons of nucleus ^{197}Pb.

nLj	$\bar{E}_{nLj}$, MeV	E_{nLj}, MeV	d	N
$1s_{\frac{1}{2}}$	-39.97	-39.93	0.67	3
$2s_{\frac{1}{2}}$	-29.56	-29.48	0.83	13
$3s_{\frac{1}{2}}$	-15.38	-15.32	0.82	17

16.5   Conclusions

Comparing the energy levels presented in Tables 16.1 and 16.2, we see that using the variation of the parameter d of the model potential, a significantly better coincidence of eigenvalues with the control calculations using a standard program EIGEN [5] has been obtained. We found that the results obtained by solving the integral equations (16.24) and (16.25) weakly depend on the parameter d or ω of the model potential for a minimum number of iterations N, if the step and limits of integration R_m were chosen correctly.

Here, we have proposed a new method with Green's functions (16.21) and derived the integral equations for negative energies (16.24) and (16.25). The expansion of perturbed eigenfunctions by the infinite number of unperturbed eigenfunctions are expressed by the series [1]

$$G(r, r') = \sum_{\alpha=0}^{\infty} \frac{U_{\alpha 0}(r) U_{\alpha 0}(r')}{E_{\alpha 0}}$$

of unperturbed eigenfunctions $U_{\alpha 0}$ and eigenvalues $E_{\alpha 0}$. We note that such expansions are very complicated for a practical solution of integral equations.

The solutions (16.25) $\varphi_\alpha U_{onL}$ at large distances are proportional to U_{0nL} or $\exp\{m\omega r^2/h\}$ and represent continuous functions. In this case, those solutions have first derivatives of the order larger than the second derivative, and therefore, they can be applied for a semi-relativistic model of atomic nuclei A[16]. In order to find relativistic corrections of eigenvalues, derivatives of the solution must be calculated till the fourth order. By using simple discretization methods, this problem can be solved only very approximately because here derivatives are defined only till the second order.

The convergence conditions of eigenfunctions are presented in (16.29) and (16.30). These results are important for practical calculations of stability of a superheavy nucleus A[16, 17] and the energy spectrum of charmed and bottom mesons A[20].

References

1. Byron, F. W. and Fuller, R. W. (1970). *Mathematics of Classical and Quantum Physics*, Dover Publications, New York.
2. Guter, R. S. and Janpolski, A. R. (1976). *Differential Equations, Vyshaya Shkola*, Moscow (in Russian).
3. Jelley, N. A. (1990). *Fundamentals of Nuclear Physics*, Cambridge University Press, New York.
4. Merzbacher, E. (1970). *Quantum Mechanics*, New York.
5. Ortega, J. M. and Pool, V. G. (1981). *An Introduction to Numerical Method for Differential Equations*, Pitman Publishing Inc., New York.
6. Petrovski, I. G. (1965). *Lectures on the Theory Integral Equations*, Nauka, Moscow (in Russian).
7. Zeldowich, J. B. (1985). Theory of perturbations for one dimension task of quantum mechanics, *J. Exp. Theoret. Phys.*, **31**, 1101 (in Russian).

Chapter 17

Potential Representation for the Coulomb Interactions

17.1 Introduction

In the potential representation, the wave function is expressed as a product of the known solution and the function which depends on the new additional potential. A new form of equations for the atomic spectroscopy were found A[11]. These equations were solved for the ground state of helium. Here, we solve a two-particle and many-particle task for bound states using the method of potential representation. In this case, the radial wave function can be represented like production of the solution without centrifugal forces on function including the centrifugal potential [1, 2]

$$f(l, \pm k, x) = \varphi(l, \pm k, x) f_0(l, \pm k, x), \tag{17.1}$$

where $\varphi(l, \pm k, x)$ is the solution depending on $V(x)$ and free solution $f_0(l, \pm k, x)$ satisfying the asymptotic condition $\lim_{x \to \infty} f_0(l, \pm k, x) \approx \exp\{\mp ikx\}$. When the potential is switched off, we require

$$\lim_{V \to 0} \varphi(l, \pm k, x) = 1. \tag{17.2}$$

At infinity for bound states,

$$\lim_{x \to \infty} f(l, \pm k, x) = 0. \tag{17.3}$$

Equity of (17.1) was proven for the Woods–Saxon potential A[1, 2] for positive energy at nuclear interactions. The aim of our investigations is the application of the potential representation method to Coulomb interactions in bound systems.

17.2 Bounded Systems: The Two-Body Task

Equation (17.1) can be true

$$\lim_{x \to \infty} f(l, \pm k, x) \approx \exp\{\mp ikx\}, \tag{17.4}$$

for bound states, when the wave number will be purely complex

$$k_{1,2} = \mp ik_n. \tag{17.5}$$

If $\varphi(l, \pm k, x)$ represents a polynomial from radial variable, then the function $f(l, \pm k, x)$ must satisfy following requirement:

$$\lim_{x \to \infty} f(l, \pm k, x) \approx x^n \exp\{-k_n x\}. \tag{17.6}$$

Using the Jost function $f_0(l, -k, x)$ [1] with the complex wave number k (17.5), it coincides with the wave function for bound states [1]. The radial Schrödinger equation

$$\frac{d^2 f}{dx^2} + \left(k^2 - \frac{l(l+1)}{x^2} - cV(x)\right) f = 0, \tag{17.7}$$

where

$$c = \frac{2m}{\hbar^2}, \tag{17.8}$$

using (17.1), it can be rewritten as follows:

$$(\varphi_x' f_0^2)_x' = V c\varphi \cdot f_0^2. \tag{17.9}$$

After integrating, we obtain

$$\varphi_x' f_0^2 = c \int V\varphi \cdot f_0^2 dx + a. \tag{17.10}$$

For $V = 0$, we have $\varphi = 1$, $\varphi_x' = 0$ and $a = 0$. Then, the definite integral can be introduced by equation

$$\varphi_x'(l, k, x) f_0^2 = \int_\infty^x cV(y)\varphi(l, k, y) f_0^2(l, k, y) dy = \Phi(x) - \Phi(\infty),$$

$$\tag{17.11}$$

which after comparing with (17.10) defines

$$a = -\Phi(\infty). \tag{17.12}$$

Finally, we obtain

$$\varphi_x'(l, -k, x) f_0^2(l, -k, x) = c \int_\infty^x V(y) \varphi(l, -k, y) f_0^2(l, -k, y) dy. \tag{17.13}$$

Inserting the free solution [4]

$$f_0\left(l + \frac{1}{2}, -k, x\right) = \sqrt{\frac{\pi}{2}} i k^{\frac{1}{2}} \exp\left\{-i\frac{\pi}{2}[l+1]\right\} H_{l+1/2}^2(-k, x) \tag{17.14}$$

in (17.13), we obtain

$$\varphi_x'(l, -k, x)(H_{l+1/2}^2(-k, x))^2$$
$$= \int_\infty^x \frac{A}{x} \varphi(l, -k, y)(H_{l+1/2}^2(-k, x))^2 dy, cV(x) = \frac{A}{x}. \tag{17.15}$$

The Hankel function of second order can be expressed by Bessel functions

$$H_{l+1/2}^2 = \frac{\exp\left\{\left(l + \frac{1}{2}\right)\pi i\right\} J_{l+1/2}(x) - J_{-l-1/2}(x)}{i \sin\left(l + \frac{1}{2}\right)\pi}, \tag{17.16}$$

which can be expanded by the power series

$$J_{l+1/2}(x) = \sum_{n=0}^\infty \frac{(-1)^n}{n!\Gamma\left(l + \frac{1}{2} + n + 1\right)} \left(\frac{x}{2}\right)^{l+\frac{1}{2}+2n}. \tag{17.17}$$

Now from (17.16) and (17.17), it follows that

$$(H_{l+1/2}^2)(-k, x))^2 = (-1)^l i \left(\sum_{n=0}^\infty d_n(2l+1)x^{2l+n+1} + \sum_{n=0}^\infty d_n(0)x^n \right.$$
$$\left. + \sum_{n=0}^\infty d_n(-2l-1)x^{-2l-n-1}\right), \tag{17.18}$$

where coefficients $d_n(2l+1)$, $d_n(0)$ and $d_n(-2l-1)$ are received by multiplying $J_{l+1/2}(-z)J_{l+1/2}(-z)$, $J_{l+1/2}(-z)J_{-l-1/2}(-z)$ and

$J_{-l-1/2}(-z)$ $J_{-l-1/2}(-z)$, $z = kx$. Considering (17.1), (17.16) and (17.17), the function φ can be presented in the following way:

$$\varphi(l, -k, x) = \sum_{n=0}^{\infty} a_n(l, -k) x^{2l+n+1}. \tag{17.19}$$

From (17.1) $f(l, \pm k, x)$ and $\lim_{x \to 0} f_0(l, \pm k, x) \approx \text{const} \cdot x^{-1}$ [2], we obtain

$$\lim_{x \to 0} f(l, \pm k, x) \approx x^{l+1}. \tag{17.20}$$

The asymptotic of $f(l, -k, x)$ at infinity when $x \to \infty$ can be obtained from $\lim_{x \to \infty} f_0(l, -k, x) \approx \exp\{ikx\}$ when $k = ik_n$

$$\lim_{x \to 0} f_0(l, -ik_n, x) \cong \exp\{-k_n x\}, \quad k_n > 0. \tag{17.21}$$

From the above formula and (2.16), we obtain

$$f(l, -ik_n, x) = \sum_{m=0}^{\infty} a_m(l, -k_n) x^{2l+m+1} e^{-k_m x}. \tag{17.22}$$

These power series must be terminated at some finite maximum power

$$a_{N+1}(l, -k_n) = a_{N+2}(l, -k_n) = 0,$$

and then, we obtain

$$\lim_{x \to \infty} f(l, -k_n, x) \approx a_{n_r} x^{2l+n_r+1} e^{-k_n x}. \tag{17.23}$$

From [4], we have the standard expression

$$\lim_{r \to \infty} U_{nl}(r) \approx e^{-\frac{Zr}{na}} \left(\frac{2Zr}{na} \right)^{l+n_r+1}. \tag{17.24}$$

which coincides with (17.23) only when $l = 0$.

The wave function can be presented in following way:

$$f(l, -k_n, x) = \varphi(l, -k_n, x) f_0 \left(\frac{1}{2}, ik_n, x \right), \tag{17.25}$$

where $f_0 \left(\frac{1}{2}, -ik_n, x \right) = e^{-k_n x}$.

The equation for the function φ

$$\varphi'_x e^{-2k_n x} = \int_\infty^x U(x)\varphi e^{-2k_n x}dx \tag{17.26}$$

follows from (17.13), where the centrifugal potential is included

$$U(x) = cV(x) + \frac{l(l+1)}{x^2}. \tag{17.27}$$

Equation (17.26) can be easily solved by expanding the functions in power series

$$\varphi(l, -ik_n, x) = \sum_{r=0}^\infty a_r(l, -ik_n)x^{l+1+r} \tag{17.28}$$

where

$$f_0\left(\frac{1}{2}, -ik_n, x\right) = \sum_{n=0}^\infty b_n x^n, \quad b_n\frac{(-2k_n)^n}{n!}. \tag{17.29}$$

Substituting (17.28) into (17.26) for the Coulomb potential, we obtain

$$\sum_{r=0}^\infty (l+r+1)a_r x^{l+r} \sum_{n=0}^\infty b_n x^n$$

$$= \int_\infty^x \left(\frac{A}{y} + \frac{l(l+1)}{y^2}\right) \sum_{n=0}^\infty a_n y^{l+n+1} \sum_{n=0}^\infty b_n y^n dy,$$

$$A = -\frac{2mZe^2}{\hbar^2} \tag{17.30}$$

Multiplying the series in a simple way [3] and considering (17.11), we obtain the final expression

$$\sum_{m=0}^{r-1} (l+m+1)a_m b_{r-l-m} - \frac{l(l+1)}{r} \sum_{m=0}^{r-1} a_m b_{r-l-m}$$

$$-\frac{A}{r} \sum_{m=0}^{r-l-1} a_m b_{r-l-m-1} = 0, \quad r = 1, 2, 3, \ldots, \quad r \geq l+1. \tag{17.31}$$

for the recursion relations between $a_r(l, -ik_n)$ for the definition of the radial wave function $f(l, -ik_n, x.)$. We can use the complicated expression (17.31) for detaching

$$a_{r-1} = \frac{1}{r(r+1) - l(l+1)} \sum_{m=0}^{r-l-1}$$

$$\times \left(A + \frac{2k(r(l+m+1) - l(l+1))}{r-l-m} \right) a_m b_{r-l-m-1}. \quad (17.32)$$

and calculating $a_n(l, -ik)$ in turn

$$a_{r-1}(l, -ik) = \prod_{m=0}^{r-l-1} \frac{A + 2k(l+m+1)}{(r-m)(r-m+1) - l(l+1)} a_0. \quad (17.33)$$

The simple formula for a_n when $l = 0$ can be easily rewritten from (17.33)

$$a_r(0, ik) = \prod_{m=0}^{r-1} \frac{A + 2k(l+m)}{(r-m)(r-m+1)} a_0. \quad (17.34)$$

Introducing $N = r - l - 1$, the last formula can be rewritten as follows:

$$a_{N+1}(l, -ik) = \prod_{m=0}^{N} \frac{A + 2k(l+m+1)}{(N+l-m+1)(N+l-m+2) - l(l+1)} \cdot a_0. \quad (17.35)$$

Now, it is easy to make a common solution

$$a_{N_r+1}(l, -ik_n) = 0, \quad k_n = -\frac{a}{2(l+N+1)}, \quad N = 0, 1, 2, \dots. \quad (17.36)$$

After substituting the expression $cV(r)$ into (17.36), we obtain [4]

$$E_n = -\frac{\hbar^2 k_n^2}{2m}, \quad E_n = -\frac{mZ^2 e^4}{2\hbar^2 n^2}, \quad n = l + N + 1. \quad (17.37)$$

Comparing the obtained result E_n with a standard equation [4], we obtain coincidence and can suppose that the considered method can be applied for more complicated systems. Eigenvalues of k_n can be found from a standard requirement for the asymptote of the wave function, and function φ satisfies the following integral differential equation:

$$\varphi'_x e^{-2k_n x} = \int_\infty^x U(x')\varphi \cdot e^{-2k_n x'}\, dx'. \tag{17.38}$$

The main aim of this part is to separate the centrifugal potential and Coulomb potential in the Schrödinger equation

$$-\frac{\hbar^2}{2m}\left(\frac{d^2 R}{dr^2} + \frac{2}{r}\frac{dR}{dr} - \frac{l(l+1)}{r^2}R\right) - \frac{Ze^2}{r}R = ER \tag{17.39}$$

by introducing the radial wave function in potential representation

$$R_{nl} = \varphi_{nl} R_{n0} \tag{17.40}$$

keeping in mind that the radial wave functions R_{n0} for $l = 0$ define the functions φ_{nl} considering that the energy E_{nl} of the hydrogen atom does not depend on l

$$R_{n0}\frac{d^2\varphi_{nl}}{dr^2} + \frac{2}{r}R_{n0}\frac{d\varphi_{nl}}{dr} + 2\frac{dR_{n0}}{dr}\frac{d\varphi_{nl}}{dr} - \frac{l(l+1)}{r^2}R_{n0}\varphi_{nl} = 0. \tag{17.41}$$

We do not need to solve this equation because it is already known. Using solutions for Coulomb potential [4], we can find φ_{nl} in (17.6) for $n = 2$, $l = 1$

$$\varphi_{21} = \frac{R_{21}}{R_{20}} = \frac{Zr}{a_0\sqrt{3}\left(2 - \frac{Zr}{a_0}\right)}. \tag{17.42}$$

Similarly, using φ_{nl} we can obtain and confirm the potential representation A[1] for any values n, l. We can affirm that for $E > 0$ A[2] and also for $E < 0$, wave functions in potential representation (17.40) can be obtained by solving the Schrödinger equation for $l = 0$ and the dependence on l can be obtained by solving (17.41). This method can be useful for potential definition of dependence on l in tasks of nuclear physics.

17.3   The Many-Particles Task

Now, we shall consider the method of potential representation and practical applications for the simple atomic Hamiltonian

$$\hat{H} = \sum_i \left(\frac{\hat{p}_i^2}{2m} - \frac{Ze^2}{r_i} \right) + \frac{1}{2} \sum_{i \neq k} \frac{e^2}{r_{ik}} \tag{17.43}$$

with a wave function similar to that in Refs. [4], A[2], i.e.,

$$\Psi(r_1, r_2, \ldots, r_n) = \varphi(r_1, r_2, \ldots, r_n) \Psi^0(r_1, r_2, \ldots, r_n), \tag{17.44}$$

where Ψ^0 is the antisymmetric wave function of the Hamiltonian $\hat{H}^0$ for the unperturbed task, where only interactions of electrons with nucleus are included

$$\hat{H}^0 = \sum_i \left(\frac{\hat{p}_i^2}{2m} - \frac{Ze^2}{r_i} \right), \tag{17.45}$$

where the symmetric function $\varphi(r_1, r_2, \ldots, r_n)$ describes the interactions between electrons. Acting with the operator $\hat{H}$ on the function Ψ, we obtain A[24]

$$\frac{1}{2m} \sum_i \left(\varphi \cdot \hat{p}_i^2 \Psi^0 + 2\hat{p}_i\varphi \cdot \hat{p}_i \Psi^0 + \Psi^0 \hat{p}_i^2 \varphi \right)$$

$$- \sum_i \frac{Ze_2}{r_i} \varphi \Psi^0 + \frac{1}{2} \sum_{i \neq k} \frac{e^2}{r_{ik}} \varphi \Psi^0 = E\varphi\Psi^0. \tag{17.46}$$

Considering the relation

$$H^0 \Psi^0 = E^0 \Psi^0, \quad E^0 = -2E_n = -2\frac{Z^2 m e^4}{2\hbar^2 n^2}, \tag{17.47}$$

Eq. (17.46) can be simplified as follows:

$$\frac{1}{2m} \sum_i (\Psi^0 \hat{p}_i^2 \varphi + 2\hat{p}_i\varphi \cdot \hat{p}_i \Psi^0) + \frac{e^2}{2} \sum_{i \neq k} \frac{1}{r_{ik}} \varphi \Psi^0 = (E - E^0)\varphi\Psi^0. \tag{17.48}$$

The obtained equation shows that by using the method of potential representation, we can share the known wave function ψ_0 and the interactions between the electrons will be included in the function φ.

17.4   Solution for the Ground State of the Helium Atom

The method solution of Eq. (17.48) can be easily explained by solving a simple task of the helium atom. The wave function Ψ for electrons having spins $s = 1/2$ must be antisymmetric. For the ground state of helium, the total spin $S = 0$ and the spin wave function $\chi(1, 2) = -\chi(2, 1)$ are antisymmetric according to the intrinsic moments or spins with different projections $s = \pm 1/2$ of two spin states as evidenced by the functions α, β. Then the radial wave function Ψ must be symmetric. The wave for two electrons of helium can be represented in the following way:

$$\Psi = \varphi(r_1)\varphi(r_2)\Psi^0(r_1)\Psi^0(r_2)\chi,$$

$$\chi(1, 2) = \frac{1}{\sqrt{2}}(\alpha(1)\beta(2) - \alpha(2)\beta(1)),$$

(17.49)

where the normalized hydrogen wave functions $\psi^0_{nlm} = R_{nl}(r) Y_{lm}(\vartheta, \varphi)$ for the spherical symmetric case $l = 0$, $m = 0$ and

$$\psi^0_{100}(r) = \Psi^0(r) = \frac{1}{\sqrt{\pi}}\frac{Z^3}{a^3} \exp\left\{-Z\frac{r}{a}\right\}$$

of two electrons was obtained [4] as follows:

$$\Psi^0(r_1, r_2) = \Psi^0(r_1)\Psi^0(r_2) = \frac{Z^3}{\pi \cdot a^3} \exp\left\{-Z\frac{r_1 + r_2}{a}\right\}, \qquad a = \frac{\hbar^2}{me^2},$$

(17.50)

where a is the Bohr radius.

After substituting (17.49) into Eq. (17.48) and averaging by r_2, we obtain two equivalent equations

$$\frac{1}{2m}(\Psi^0(r_1)\hat{p}_i^2\varphi(r_1) + 2\hat{p}_i\Psi^0(r_1)\hat{p}_i\varphi(r_1))$$

$$+ \frac{1}{2}\left\langle\frac{e^2}{r_{12}}\right\rangle \varphi(r_1)\Psi^0(r_1) = \frac{E - 2E^0}{2}\varphi(r_1)\Psi^0(r_1). \quad (17.51)$$

We obtain the second equation by replacing the variable r_1 with r_2. We obtained the equation which shows that the variables cannot be separated. Equation (17.51) can be solved by considering the interaction of one of the electrons with the average field of another A[11], i.e.,

$$\frac{1}{2m}(\Psi^0(r_1)\hat{p}_i^2\varphi(r_1) + 2\hat{p}_i\Psi^0(r_1)\hat{p}_1\varphi(r_1))$$

$$+\frac{1}{2}\int\frac{e^2}{r_{12}}|\varphi(r_2)\Psi^0(r_2)|^2dV_2\varphi(r_1)\Psi^0(r_2)$$

$$=\frac{E-2E^0}{2}\varphi(r_1)\Psi^0(r_1). \qquad (17.52)$$

For the first approach $\varphi(r_2) = 1$ and solving (17.52), we find $\varphi(r_1)$ and more exactly $\varphi(r_2)$. We can repeat these procedures till both the functions $\varphi(r_1)$ and $\varphi(r_2)$ coincide. But we solve (17.52) using analytical method which is similar [4] to the variation according to the parameter Z. Calculating the integral in equation (17.52) and taking $\varphi(r_2) = 1$, we obtain

$$4\pi\int_0^\infty|\psi^0|^2\frac{1}{r_{12}}r_2^2dr_2 = -\frac{Z}{a}e^{-2Z\cdot r_1/a} - \frac{1}{r_1}e^{-2Z\cdot r_1/a} + \frac{1}{r_1}. \qquad (17.53)$$

Limiting the last equation for the case $r_1 \to \infty$ only by the third term, we obtain

$$\varphi(r_1) = \text{Const}\exp\{\alpha\cdot r_1\}, \qquad (17.54)$$

$\alpha = \frac{1}{2a} > 0$, with double ionization energy, having lower approximate value $E = 1.125E^0$. The function φ is exponentially increasing, similar to the variation by the parameter Z. From (17.50) and (17.54), one-particle wave function in potential representation is given as follows:

$$\Psi(r_1) = \frac{(Z-\beta)^{\frac{3}{2}}}{\sqrt{\pi}a^{\frac{3}{2}}}\exp\left\{\beta\frac{r_1}{a}\right\}\exp\left\{-Z\frac{r_1}{a}\right\}. \qquad (17.55)$$

Calculating the average values of terms of Eq. (17.53), using function (17.55) we obtain

$$B^2 + 2B(2A-B) - 4AB - \frac{5}{8}A(2A-B) = \frac{k^2}{2} - k_0^2, \qquad (17.56)$$

where

$$B = \frac{\beta}{a}, \quad \frac{1}{a} = A, \quad k_0^2 = 4A^2 \qquad (17.57)$$

Requiring that energy of the ground state must have maximum value

$$\frac{\partial k^2}{\partial B} = 0, \tag{17.58}$$

from (17.56), we obtain

$$B = 0.3125A. \tag{17.59}$$

Substituting the obtained value into (17.56), the following formula was obtained:

$$\frac{k^2}{2} = 2.847A^2. \tag{17.60}$$

Here, A is wave number fit to the bound energy of the electrons in the hydrogen energy A[11], i.e.,

$$E = -2.847\frac{me^4}{\hbar^2} = -77.41\,\text{eV}. \tag{17.61}$$

This result practically coincides with the result $E = -77.49\,\text{eV}$ of the Hartree–Fock calculations [5]. The difference of our result from the experimental value $E_{eksp.} = -79.02\,\text{eV}$ can be in the approximate function (17.55) of φ without solving (17.52). The obtained equations (17.48) differ from the Hartree–Fock equations. The potential representation method can be useful for atomic and molecular spectroscopy and for allotting central forces field in the nuclei calculations. The calculations of quantum equations can be simplified using the potential representation method and function φ depending on share of the potential.

17.5 Variation According to the Parameter Z

For the first-order approximation of bounding energy of electrons in the He atom, we must calculate the matrix element of interaction energy between electrons using the wave function (17.50)

$$\Delta E_{in} = \iint \Psi^{0*}(r_1, r_2)\frac{e^2}{r_{12}}\Psi^0(r_1, r_2)d\tau_1 d\tau_2, \tag{17.62}$$

and expression

$$\frac{1}{r_{12}} = \frac{4\pi}{r_>} \sum_{l,m} \frac{1}{2l+1} \left(\frac{r_<}{r_>}\right)^l Y^*_{lm}(\vartheta_1, \varphi_1) Y_{lm}(\vartheta_2, \varphi_2). \qquad (17.63)$$

Substituting (17.63) into (17.62), we obtain the integrals

$$\Delta E_{in} = \frac{4e^2}{\pi} \left(\frac{Z}{a}\right)^6 \int_0^\infty e^{-\frac{2Z}{a} r_1}$$

$$\times \left[\frac{1}{r_1} \int_0^{r_1} e^{-\frac{2Z}{a} r_2} r_2^2 r_2 + \int_{r_1}^\infty e^{-\frac{2Z}{a} r_2} r_2 dr_2\right] r_1^2 dr_1 \qquad (17.64)$$

that define the repulsion energy between two electrons as follows:

$$\Delta E_{in} = 5\frac{Ze^4}{8a} = \frac{5}{8}\frac{Zme^4}{\hbar^2}. \qquad (17.65)$$

Then the energy of the ground state of helium atom of configuration $1s^2$ for state $^1S_0$ or $^{2s+1}S_j$ was obtained [4] from (17.47) and (17.65) as follows:

$$E_{\text{He}} = E^0 + \Delta E_{in} = -\frac{Z^2 me^4}{\hbar^2 n^2} + \frac{5}{8}\frac{Zme^4}{\hbar^2}, \quad n = 1, \ Z = 2. \qquad (17.66)$$

Now requiring the energy minimum according to the variation of parameter Z, we have

$$\frac{\partial E_{\text{He}}}{\partial Z} = 0, \quad Z_e = 27/16 = 1.687, \qquad (17.67)$$

and the following value of energy [6] was obtained:

$$E_{\text{He}} = -2.904 \frac{e^2}{a} = -2.904 \frac{me^4}{\hbar^2} = -78.96 \,\text{eV},$$

$$\frac{me^4}{\hbar^2} = 27.19 \,\text{eV}, \qquad (17.68)$$

which differs from the energy of the full ionization of the helium atom by only 1.9%. The obtained result is more exact than (17.61), which was calculated by considering $\varphi(r_2) = 1$.

References

1. Baz, A. I., Zeldowich, J. B. and Perelomov, A. M. (1966). *Scattering, Reactions and Fissions in Nonrelativistic Quantum Mechanics*, Moscow, Nauka, p. 339.
2. Alfaro, V. and de Redge, G. (1966). *Potential Scattering*, Moscow, Mir, p. 274.
3. Lavrentjev, A. M. and Shabat, B. M. (1965). *Theory Complex Variable*, Moscow, State Publishing, p. 716.
4. Shiff, L. (1959). *Quantum Mechanics*, Moscow, IL, p. 473.
5. K. Way (ed.) (1974). *Atomic Data and Nuclear Data Tables*, New York and London, Academic Press, p. 478.

Chapter 18

Transformations of the Hamiltonian for Jastrow's Correlation Method

18.1 Introduction

Using the potential representation method, we transformed the Schrödinger equation to the two related differential equations. In the first step, we must solve the Schrödinger equation or use the known solution for the one-particle approach in the central field. Then, we must solve a differential equation for finding a function, which depends on the interaction potential between particles A[1]. Using this method, a new form of the Schrödinger equation for the atomic spectroscopy was found. This equation was solved for the ground state of helium.

The main idea of the potential representation method is that if we add to the Hamiltonian an additional potential, we can obtain the new solution multiplying the previous solution by a function, which depends on the additional potential. This method was proposed in the paper A[2].

The aim of this chapter is to obtain and solve a differential equation for a function, which depends on the interaction potentials between electrons e^2/r_{ik}. The analytical solutions for a hydrogen-type atom, where the potentials between the electron and the nucleus with the charge Ze are included, are obtained A[12].

Now, we will consider the possibility to adapt this method to the atomic spectroscopy calculation method when the Hamiltonian is (see [1]) given as follows:

$$\hat{H} = \sum_i \left(\frac{\vec{p}_i^2}{2m} - \frac{Ze^2}{r_i} \right) + \frac{1}{2} \sum_{i \neq k} \frac{e^2}{r_{ik}},$$

$$\vec{p}_i^2 = -\hbar^2 \Delta_i. \tag{18.1}$$

Here, Δ_i is the Laplacian operator. The first term on the right-hand side of (18.1) is the Hamiltonian for electrons of the hydrogen-type atom, and the second term is the interaction potential between the electrons.

According to the potential representation method [2], the wave function for atoms can be written in the following form:

$$\Psi(\vec{r}_1, \vec{r}_2, \ldots, \vec{r}_n) = \varphi(\vec{r}_1, \vec{r}_2, \ldots, \vec{r}_n) \cdot \Psi^0(\vec{r}_1, \vec{r}_2, \ldots, \vec{r}_n), \tag{18.2}$$

where Ψ^0 is the antisymmetrical wave function according to changing the coordinates of any pair of the electrons. The function Ψ^0 is an eigenfunction of the unperturbed Hamiltonian

$$\hat{H}_0 = \sum_i \left(\frac{\vec{p}_i^2}{2m} - \frac{Ze^2}{r_i} \right). \tag{18.3}$$

$\varphi(\vec{r}_1, \vec{r}_2, \ldots, \vec{r}_n)$ is a symmetrical function which depends on the repulsion potential between electrons

$$V_{ee} = \sum_i \frac{1}{2} \sum_{i \neq k} \frac{e^2}{r_{ik}}. \tag{18.4}$$

The aim of this chapter is to prove the following result.

Theorem. *The differential equation for finding the function φ for the ground state of He atom exists.*

If we apply the Hamiltonian (18.1) with potential (18.4) to the wave function (18.2) and use the relation

$$\hat{H}_0 \Psi^0 = E_0 \Psi^0, \tag{18.5}$$

we obtain the following equation:

$$\frac{1}{2m} \sum_i (\Psi^0 \hat{p}_i^2 \varphi + 2\hat{p}_i \varphi \cdot \hat{p}_i \Psi^0) + \frac{e^2}{2} \sum_i \frac{1}{r_{ik}} \varphi \Psi^0 = (E - E_0) \varphi \Psi^0. \tag{18.6}$$

The obtained equation shows that, using the method of potential representation, we can separate the interaction between electrons and the interaction between electrons and nucleus. The function φ contains information about the expectation value of interaction between electrons V_{ee}.

Considering the ground states of closed s–d shells of nuclei, we can use the Jastrow wave function [2]

$$\Psi(\vec{r}_1, \vec{r}_2, \ldots, \vec{r}_A) = (C_A)^{-\frac{1}{2}} \prod_{1 \leq i \langle j \leq A} f(r_{ij}) \Phi(\vec{r}_1, \vec{r}_2, \ldots, \vec{r}_A), \quad (18.7)$$

for A fermions. Here, C_A is a normalization constant, and $r_{ij} = |\vec{r}_i - \vec{r}_j|$. Φ is a single Slater determinant [1]. Comparing the last expression with (18.2), we see that (18.7) is a similar approximation. For the ground state of helium, according to (18.2) and (18.7), we will use the following expressions for the radial wave functions [3]:

$$\Psi(r_1, r_2) = C\varphi(r_{12}) \cdot R_{10}(r_1) R_{10}(r_2),$$

$$\Psi^0 = R_{10}(r_1) R_{10}(r_2) = \frac{\gamma^3}{\pi} \cdot \exp\{-\gamma(r_1 + r_2)\},$$

$$\gamma = \frac{Z}{a_0}, \quad (18.8)$$

$$a_0 = \frac{\hbar}{me^2},$$

where C is a normalization constant, a_0 denotes the first Bohr radius of hydrogen if m is the mass and e is the charge of the electron.

18.2 Transformation of the Hamiltonian for He Atom

Using the Jastrow approach (18.8), we can transform the Hamiltonian for the He atom. Applying Laplaces' operators Δ_1 and Δ_2, acting on r_1 and r_2 to $a = \varphi(r_{12})R(r_1)$ and $b = \varphi(r_{12})R(r_2)$, we obtain the following equations:

$$\Delta_1 a = \varphi \frac{d^2 R(r_1)}{dr_1^2} + \frac{2}{r_1} \varphi \frac{dR(r_2)}{dr_1} + R(r_1) \frac{d^2\varphi}{dr_{12}^2}$$

$$+ \frac{2}{r_{12}} R(r_1) \frac{d\varphi}{dr_{12}} + 2 \frac{dR(r_1)}{dr_1} \cdot \frac{d\varphi}{dr_{12}} \cos\alpha, \quad (18.9)$$

$$\Delta_2 b = \varphi \frac{d^2 R(r_2)}{dr_2^2} + \frac{2}{r_2} \varphi \frac{dR(r_2)}{dr_2} + R(r_2) \frac{d^2 \varphi}{dr_{12}^2}$$

$$+ \frac{2}{r_{12}} R(r_2) \frac{d\varphi}{dr_{12}} + 2 \frac{dR(r_2)}{dr_2} \cdot \frac{d\varphi}{dr_{12}} \cos \beta. \qquad (18.10)$$

The angles α, β and ϑ_{12} are presented in the figure below. Considering that ends of the vectors $\vec{r}_1$ and $\vec{r}_2$ have the spherical coordinates $r_1, \vartheta_1, \varphi_1$ and $r_2, \vartheta_2, \varphi_2$, respectively, we obtain

$$\cos \alpha = \frac{-\vec{r}_1 \cdot \vec{r}_{12}}{r_1 \cdot r_{12}} = \frac{r_1 - r_2 \cos \vartheta_{12}}{r_{12}},$$

$$\cos \beta = \frac{-\vec{r}_2 \cdot \vec{r}_{12}}{r_2 \cdot r_{12}} = \frac{r_2 - r_1 \cos \vartheta_{12}}{r_{12}}. \qquad (18.11)$$

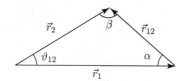

The wave function in the zero approximation for the ground state of He atom is $\Psi^0 = R_{10}(r_1)R_{10}(r_2)$ and does not depend on the spherical angles ϑ, φ (here, we have the quantum numbers $n = 1$, $l = 0$). From (18.3) and (18.5), we obtain $E_0 = 2E_{10}$. In this approach, where only the interaction between electrons and nucleus with $Z = 2$ are included, we can obtain E_0 using the energy expression E_n of atom with one electron

$$E_n = -\frac{Z^2 m e^4}{2\hbar^2 n^2},$$

$$E_0 = 2E_{n=1},$$

$$E_0 = -2\frac{Z^2 m e^4}{2\hbar^2} \qquad (18.12)$$

Acting with the Hamiltonian (18.1) on the wave function Ψ (18.8) and using (18.5) and (18.9)–(18.11), we find

$$-\frac{\hbar^2}{2m}\left(2\frac{d^2\varphi}{dr_{12}^2} + \frac{4}{r_{12}}\frac{d\varphi}{dr_{12}} + 2\sum_{i\neq j}^{2}\frac{r_i - r_j\cos\vartheta_{12}}{r_{12}}\frac{\frac{dR_{10}(r_i)}{dr_i}}{R_{10}(r_i)}\frac{d\varphi}{dr_{12}}\right)$$

$$+\frac{e^2}{r_{12}}\varphi = \Delta E\varphi, \tag{18.13}$$

where the ionization energy E of two electrons of He atom in the ground state according to (18.6) can be expressed by

$$E = E_0(Z = 2) + \Delta E. \tag{18.14}$$

This leads to the differential equation for φ and the theorem is proved. Since the expectation value $\langle\cos\vartheta_{12}\rangle$ for the ground state of He is given as

$$\langle\cos\vartheta_{12}\rangle = 2\pi\int_0^\pi \cos\vartheta_{12}\cdot\sin\vartheta_{12}d\vartheta_{12} = 0, \tag{18.15}$$

replacing r_1, r_2 in (18.13) by the expectation values $\langle r_1\rangle$ and $\langle r_2\rangle$ (see [1]), and using (18.8), we obtain

$$\langle r_1\rangle = \langle r_2\rangle = \frac{3}{2Z}a_0. \tag{18.16}$$

We can solve the modified equation (18.13)

$$-\frac{\hbar^2}{2m}\left(2\frac{d^2\varphi}{dr_{12}^2} + \frac{4 - 4\cdot\frac{3}{2Z}a_0\gamma}{r_{12}}\frac{d\varphi}{dr_{12}}\right) + \frac{e^2}{r_{12}}\varphi = \Delta E\varphi \tag{18.17}$$

analytically. We can obtain the real physical solution of this equation at infinity in the form of a real-valued function

$$\varphi = \text{const}\cdot\exp\left\{-r_{12}\sqrt{\frac{m\Delta E_t}{\hbar^2}}\right\} \tag{18.18}$$

if $\Delta E = -\Delta E_t < 0$, only.

When the perturbation potential represents the repulsion between the electrons, the positive term $2\Delta E_t$ must be added to the modified unperturbed eigenvalue E_0. This repulsion potential increases the energy E of the He atom by a positive term ΔE_t [3]. In this case, the wave function Ψ^0 for the unperturbed Hamiltonian H_0 must be changed. The solution of Eq. (18.13) in the case $\Delta E = -\Delta E_t$ can be expressed in the following form:

$$\varphi_m = C \cdot r_{12}^2 \cdot \exp\left\{-r_{12}\sqrt{\frac{m\Delta E_t}{\hbar^2}}\right\},$$

$$\Delta E_t = \frac{me^4}{\hbar^2}.$$

$$(18.19)$$

Now the eigenvalue E_{0t} for the modified unperturbed solution (18.5) is

$$E_{ot} = 2E_{10} + 2\Delta E_t = -2 \cdot \frac{me^2}{\hbar^2}. \qquad (18.20)$$

Taking into account (18.11) and substituting $Z = \sqrt{2}$ in (18.8), we can get the modified unperturbed wave function Ψ_t^0. From (18.11), (18.15) and (18.16) we have

$$E = E_{0t} + \Delta E = -3 \cdot \frac{me^4}{\hbar^2}, \quad E = -81.57\,\text{eV}. \qquad (18.21)$$

The obtained result (18.21) is in good coincidence with experiments (see [1]): $E_{ex} = -2{,}904\frac{me^4}{\hbar^2}$. To obtain a more exact result, we must find new expectation values $\langle r_1 \rangle$ and $\langle r_2 \rangle$. Instead of the wave functions (18.8) and (18.19), the expression

$$\Psi_t(r_1, r_2 \cdot r_{12}) = \varphi(r_{12}) \cdot \Psi_t^0(r_1, r_2) \qquad (18.22)$$

must be used and then Eqs. (18.13) and (18.14) must be solved. Applying this procedure, step by step, we can obtain an analytical expression for the wave function with desirable accuracy.

18.3   Conclusions

The proposed method also enables us to get approximate analytical solutions for any state of any atom. However, for the transformations (18.9) and (18.10), more complicated wave functions must be used. The proposed mathematical model also can be used for the transformation of a single-particle shell model to many-particle model of nuclei, where the known solutions Ψ^0 and energies E_0 can be used similar to the initial approach for including the interactions of nucleons in the shells.

References

1. Bandzaitis, A. and Grabauskas, D. (1975). *Quantum Mechanics*, Mokslas, Vilnius, p. 340 (in Lithuanian).
2. Stoitsov, M. V., Antonov, A. N. and Dimitrova, S. S. (1993). Natural orbital representation and short-range correlations in nuclei, *Phys. Rev. C*, **48**(1), 74–86.
3. Schiff, L. I. (1955). *Quantum Mechanics*, McGraw-Hill, New York, p. 473.

Chapter 19

Stability of Nuclei

We can consider the static properties and radioactive decay of nuclei when relativistic corrections for the masses of nucleons are included A[14, 15]. These corrections are increasing the binding energies of the neutrons and protons in the external shells and take part in the stability of superheavy nuclei A[16, 17] with the enlarging role of shell effects [1]. In Ref. [2], the experimental method for the production of neutron-rich ($N = 184$) superheavy nuclei by low-energy multinucleon transfer reactions using shell effects like in $^{136}_{54}$Xe collisions with $^{126}_{82}$Pb was proposed. Here, the stabilization effect of closed shells A[14] and the significant relativistic effects for masses A[14] of neutrons and protons by quantum diffusion A[21] in a large region by the excited states of projectiles and targets can be used. The relativistic corrections to the potential are positive and small and can compensate the negative corrections for the mass of light nuclei A[15]. The relativistic corrections for nucleon masses are very important for the definition of heavy and superheavy nuclei energy levels of external shells A[16]. The relativistic corrections of the mass depend on the principal and the orbital quantum numbers and do not have any influence on the definition of the constant of the spin–orbit interaction. The negative corrections for the mass or energy levels for protons and neutrons depend on nuclei and vary approximately from -0.2 MeV to -1 MeV. It is important for the interpretation of α- (alpha), β- (beta) and γ- (gamma) decays. The number of decays occurring during the time interval from t to $(t + \Delta t)$

$$\Delta N = -\lambda N_t \Delta t, \tag{19.1}$$

is proportional to the number of parent nuclides N_t and the decay or disintegration constant λ. Rewriting (19.1) like a differential equation, after integration we obtain the radioactive decay law $N_t = N_0 e^{-\lambda \cdot t}$,

$$\lambda = -\frac{\ln(N_t/N_0)}{t} = -\frac{\ln(1/2)}{T_{1/2}} = \frac{0.693147}{T_{1/2}}, \tag{19.2}$$

defining the number N_t of non-disintegrated nuclei after time t when N_0 is the number of parent nuclides at $t = 0$. The half-life time $T_{1/2}$ is defined as the time it takes for half of a given number of radioactive nuclei to decay. In α decay, ${}_2^4\mathrm{He}$ is formed by the strong attractive potential between the neutrons and protons. ${}_2^4\mathrm{He}$ must penetrate the positive repulsive Coulomb barrier at the surface of the nucleus and like other β and γ decays, the electron capture and gamma decay have a stochastic character. The pathways by which a radioactive nucleus can decay can be represented [3] in the following way:

$$_Z^A X \rightarrow {}_{Z-2}^{A-4} X + {}_2^4\mathrm{He}, \tag{19.3}$$

$$_Z^A X \rightarrow {}_{Z+1}^{A} X + \beta^- + \bar{v}, \tag{19.4}$$

$$_Z^A X \rightarrow {}_{Z-1}^{A} X + \beta^+ + v, \tag{19.5}$$

$$_Z^A X + {}_{-1}^{0} e \rightarrow {}_{Z-1}^{A} X + v, \tag{19.6}$$

$$_Z^A X^* \rightarrow {}_{Z+1}^{A} X + \gamma. \tag{19.7}$$

Disintegration energies of the alpha decay [3, 4] for unstable nuclei

$$\mathcal{Q}_\alpha = (M_X - M_Y - M_\alpha)c^2 = \Delta M \cdot c^2, \quad \Delta M > 0 \tag{19.8}$$

can be defined by the masses of the parent M_X and the daughter M_Y nuclei and the mass of the alpha particle M_α. The residual or disintegration energy $\mathcal{Q}_\alpha$ appears in the form of kinetic energy of the daughter nucleus and alpha particle ${}_2^4\mathrm{He}$.

Table 19.1.   Masses of the atoms elementary particles;
u is the unified atomic mass unit.

Particle	kg	u	MeV/c^2
Proton	$1.6726 \cdot 10^{-27}$	1.007276	938.28
Neutron	$1.6750 \cdot 10^{-27}$	1.008665	939.57
Electron	$9.109 \cdot 10^{-27}$	$5.486 \cdot 10^{-4}$	0.511

The stability of heavy nuclei [1, 4] depends on the binding energy of the α particle, which is 28.3 MeV. The binding energy of a nucleon in the external shells of the heavy nuclei is about 7 MeV. For removing two protons and two neutrons from a nucleus with a binding energy of 28.3 MeV, changes in the binding energies in the structure of the parent nucleus must be used. Alpha decays in this case can define the stability [6] of the heavy nuclei A[16, 17]. For the calculations of decays and reactions, we present the rest mass of the proton m_p, neutron m_n and electron m_e in various units [3] in Table 19.1.

Here, the unified mass unit $u = 1.660559 \cdot 10^{-27}$ kg $= 931.50$ MeV/c^2 is presented. The difference in mass Δm between the separate nucleons and nucleus containing these nucleons multiplied by c^2 defines the binding energy of the nucleus, i.e.,

$$E_b(\text{MeV}) = [Zm_H + Nm_n - M(^A_Z X)] \cdot 931.50 \text{ MeV}/u = \Delta m \cdot c^2.$$
$$(19.9)$$

β^- decay (19.4) of a nucleus to another nucleus with the same mass number A occurs when a neutron in this parent nucleus disintegrates to a proton with the emission of an electron and an anti-neutrino:

$$n \rightarrow p + e^- + \bar{v}. \qquad (19.10)$$

Similarly, β^+ decay (19.5) is to the decay of a proton to a neutron, positron e^- and neutrino v. Electron capturing occurs when the proton in a nucleus changes to a bound neutron interacting with one of the atomic electrons and emitting a neutrino. The conservation of energy requires that a nucleus in the beginning state must be heavier than the mass of the final nucleus and the particles produced after the decay. The conditions for β^-, β^+ decay to be energetically possible [4] are as follows:

$$\begin{aligned}^A_Z M > ^A_{Z+1} M, \end{aligned} \tag{19.11}$$

$$\begin{aligned}^A_Z M > ^A_{Z-1} M + 2m_e, \end{aligned} \tag{19.12}$$

$$\begin{aligned}^A_Z M > ^A_{Z-1} M. \end{aligned} \tag{19.13}$$

The β^-, β^+ decays happen when quarks u and d, which consist of neutrons $n(udd)$ and protons $p(udd)$, are exchanged by intermediate vector bosons W interacting in the following way:

$$d \to u + W^-, \quad n \to p + e^- + \bar{\nu}_e, \tag{19.14}$$

$$u \to d + W^+, \quad p \to n + e^+ + \nu_e. \tag{19.15}$$

The range of their reactions is equal to the distance by spreading mediators of the virtual bosons $W^\pm$

$$R = c\Delta t, \tag{19.16}$$

and existence time of which can be defined by uncertainty relation

$$\Delta t = \hbar/\Delta E = \hbar/mc^2, \quad R_W = \hbar/mc \approx 10^{-16}\,\text{cm},$$
$$m = m_W = 80\,\text{GeV}/c^2. \tag{19.17}$$

The range $R_\pi \approx 1.4 \cdot 10^{-13}\,\text{cm} = 1.4\,\text{fm}$ of the strong interaction between two neutrons can be calculated (19.17) using the mass $m_\pi = 140\,\text{MeV}/c^2$ of the generated virtual pion taking part in exchanges. The result R_W can be obtained by diffusion of quarks in physical vacuum [1] by interaction with virtual intermediate bosons.

The produced nucleus has a greater binding energy than the initial decaying nucleus. The binding energies determine which decays are energetically possible. The spontaneous binary fission of uranium light isotopes $^{235}_{92}\text{U}$, $^{233}_{92}\text{U}$ divides the two nucleus into two roughly equal-sized nuclei, emitting 2 or 3 neutrons and generating a chain reaction with the release of nuclear energy.

A complete list of nuclides of the elements [5] up to protons $Z = 102$ comprises about 1050 nuclides, out of which 25% are stable. The band of stability stops at element $^{209}_{83}\text{Bi}$ for $Z = 83$ when $N/Z = 1.52$.

We must remark that more neutrons are needed for compensation of the increasing proton–proton repulsion. For many light nuclei, such as $^2_1$H, $^4_2$He, $^{12}_6$C, $^{14}_7$N, $^{16}_8$O, we have approximately $N \approx Z$, $N = A - Z$.

A very important role is played by the shell effects and relativistic corrections to masses of nucleons for the stability of the nucleus A[16, 17]. As a result of shell model and experiments, it has been found that nuclei with even numbers of protons and neutrons are more stable [4] than those with odd numbers. The most stable nuclei are those with the magic numbers [1] of protons and neutrons

$$Z = 2, 8, 20, (28, 40)50, 82, 114, (126), \text{ or } 164, \text{ and} \quad (19.18)$$

$$N = 2, 8, 20, (28, 50)82, 126, 184, 196(272), 318, \quad (19.19)$$

having most binding energies in external shells. The existence of magic numbers of nucleons shows that nuclei have the internal structure similar to the atomic shell model. This requires the existence of an average spherically symmetric field and that the shell energies can be calculated using the Woods–Saxon potential. We obtained that the relativistic corrections to the nucleons masses are increasing the bound energies A[15] and stability of the nuclei A[16, 17]. The calculated one-nucleon energy levels for hypothetical A[16] nucleus $^{340}_{126}X$ admit the following magic numbers for the superheavy

$$Z = 114, 120, 124, 126, 138, \text{ and} \quad (19.20)$$

$$N = 184, 196, 214 \quad (19.21)$$

and other nuclei (19.18), (19.19).

The number of the nucleons in the energy levels is limited by the Pauli's principle where the order of the level is defined by the spin–orbit interaction. Our aim is to find realistic magic numbers for the superheavy nuclei and calculate the meanings of relativistic effects in the external magic shells of neutrons and protons, to evaluate the stability of the nuclei and apply semi-relativistic theory to elementary particles. It is interesting to remark that our A[16] theoretical results (19.20) and (19.21) coincide with the calculated [6] spherical doubly magic nuclei $^{298}_{114}X$, $^{292}_{120}X$ and $^{310}_{126}X$, where the relativistic, complicated mean field model having a large number of parameterizations of interactions was used. According our calculations, A[16] the most stable must be the doubly magic nucleus $^{298}_{114}X$, $^{292}_{120}X$ and $^{310}_{126}X$ because their relations N/Z are, respectively 1.61, 1.78 and 170 and the disintegration energies are $Q_\alpha < 0$. But these important relations between the numbers of neutrons and protons can be studied experimentally [7]. Here, we investigated the reactions of $^{197}_{79}Au$ with the target $^{232}_{90}Th$, where nuclei with $Z > 100$ can be produced by heavy mass transfer for different energies of colliding nucleus excitations. The idea of detection of the superheavy nucleus produced is based [7] on the characteristic alpha decay.

References

1. Devanarayan, S. (2016). *A Text Book on Nuclear Physics*, Lexington, USA.
2. Valery Z. and Walter, G. (2010). New ideas on the production of heavy and superheavy neutron rich nuclei, *Nucl. Phys. A*, **834**, 366–369.
3. Serway, R. A. (1990). *Physics for Scientists & Engineers with Modern Physics*, Saunders College Publishing, Philadelphia, p. 1444.
4. Jelley, N. A. (1990). *Fundamentals of Nuclear Physics*, Cambridge University Press, New York, p. 278.
5. Enge, H. A. (1966). *Introduction to Nuclear Physics*. Addison Wesley Publishing Company, INC, USA, p. 582.
6. Bender, M., Rutz, K. *et al.* (1999). Shell structure of superheavy nuclei in self-consistent mean-field models, *Phys. Rev. C*, **60**, 034304-1.
7. Wieloch, A. *et al.* (2016). A novel approach to the island of stability of superheavy elements search, *EPJ* (Web of conferences), **117**, 01003.

Chapter 20

Relativistic Corrections for Neutrons in the Harmonic Oscillator Well

20.1 Introduction

Usually, nuclei are considered like non-relativistic systems. This assumption is based on the fact that the binding energy of a nucleon is significantly smaller than the energy of its rest mass. However, the strong repulsive forces at small distances and the significant spin–orbit interaction require that the relativistic corrections for mass and potential must be included [1]. Though the binding energies of the nucleons compose only 1% of the rest energy, the decreasing expectation value of the kinetic energy that is obtained is a significant value. In nuclear physics, a semi-relativistic Hamiltonian [2] must be used

$$\hat{H} = \frac{\hat{p}^2}{2m} + V(r) + \frac{1}{2m^2c^2}\frac{1}{r}\frac{dV(r)}{dr}(\vec{s}\vec{l}) - \frac{\hat{p}^4}{8m^3c^2} - \frac{\hbar^2}{4m^2c^2}\frac{dV(r)}{dr}\frac{d}{dr}.$$

(20.1)

The fourth $\hat{H}_m$ and fifth $\hat{H}_V$ terms in the $\hat{H}$ expression represent the relativistic corrections for the mass and the potential to the non-relativistic Hamiltonian $\hat{H}_0$

$$\hat{H}_p = -\frac{\hat{p}^4}{8m^3c^2} - \frac{\hbar^2}{4m^2c^2}\frac{dV(r)}{dr}\frac{d}{dr} = H_m + H_V,$$

$$\hat{H}_0 = \frac{\hat{p}^2}{2m} + V(r),$$

(20.2)

$$H - m_0c^2 = \sqrt{m_0^2c^4 - p^2c^2} = +\frac{p^2}{2m_0} - \frac{p^4}{8m_0^3c^2}.$$

(20.3)

The third term in (20.1) represents the spin–orbit interaction thus providing evidence for a repulsive core influence on the spin–orbit component of the nucleon–nucleon potential in the shell model [2], and by the meson exchange theory, we have

$$V_{sl}(r) = \lambda \frac{1}{2m^2c^2} \frac{1}{r} \frac{dV(r)}{dr} (\vec{s}\vec{l}). \tag{20.4}$$

In this chapter, the relativistic corrections for mass of neutrons and potential are considered for harmonic oscillator potential, i.e.,

$$V(r) = \frac{m\omega^2 r^2}{2} - V_0. \tag{20.5}$$

20.2   The Semi-relativistic Hamiltonian

In order to acheive this, we must use a concrete expression for

$$\hat{p}^4 = 2m\hat{T} \cdot 2m\hat{T}, \tag{20.6}$$

and kinetic energy

$$\hat{T} = -\frac{\hbar^2}{2m} \frac{1}{r^2} \frac{d}{dr} \left(r^2 \frac{d}{dr} \right) + \frac{\hat{L}^2}{2mr^2} \tag{20.7}$$

From the above equations, we obtain the following expression:

$$\hat{p}^4 = \hbar^4 \left(\frac{d^4}{dr^4} + \frac{4}{r} \frac{d^3}{dr^3} - \frac{\hat{L}^2}{\hbar^2} \left(\frac{2}{r^2} \frac{d^2}{dr^2} + \frac{2}{r^4} \right) + \frac{\hat{L}^4}{\hbar^4} \frac{1}{r^4} \right), \tag{20.8}$$

which can be introduced to an operator of relativistic correction (20.2) $\hat{H}_p$ for nucleon masses and potentials in order to obtain the expression for practical calculations of relativistic corrections for the first approach

$$\Delta E_p = \int_0^\infty \varphi_{nl}^*(r) \hat{H}_p \varphi_{nl}(r) r^2 dr, \tag{20.9}$$

using radial wave functions $\varphi_{nl}(r)$ for the harmonic oscillator (20.5). The energies E_{nlj} of quantum states nlj, with the different quantum numbers j and nlj, are splitting. The levels with different moments j_1 and j_2 for neutrons, in the one-particle shell model, with the spin–orbit interaction are increasing the binding energies for orbital quantum numbers l and the large moment j_1, i.e.,

$$E_{Nlj} = -V_0 + \left(N + \frac{3}{2}\right)\hbar\omega - a_{lj}m\omega^2;$$

$$a_{ij} = \frac{\lambda}{2}\left(\frac{\hbar}{mc}\right)^2 \begin{cases} l, & j_1 = l + \frac{1}{2}, \\ -(l+1), & j_2 = l - \frac{1}{2}; \end{cases} \tag{20.10}$$

$$N = 2n + 1 - 2, \quad N = 0, 1, 2, \ldots, \quad n = 1, 2, 3, \ldots.$$

20.3   Results and Conclusions

V_0 and ω were defined from the energy levels $2s$ and $3s$ of neutrons in $_{79}\mathrm{Au}^{197}$ the nucleus A[13]:

$$\omega = 1{,}0903 \cdot 10^{22}\,\frac{\mathrm{rad}}{c}; \quad V_0 = 56{,}327\,\mathrm{MeV} \tag{20.11}$$

$$\lambda = 16{,}53 \tag{20.12}$$

We defined the spin–orbit constant λ from the energy levels $1p_{3/2}$ and $1p_{1/2}$. For these parameters some energy levels of neutrons in $_{79}\mathrm{Au}^{197}$ coincide with the results [4, 5] for the Woods–Saxon potentials. The most frequently calculated single-particle energy levels E_{nlj} for a harmonic oscillator potential (20.10) and (20.9) of neutrons, spin–orbit interactions E_{ls} and relativistic corrections ΔE_p are presented in Table 20.1.

Relativistic corrections ΔE_p have a negative sign and shown an increase in value for greater quantum numbers n and l. These results are connected with the average kinetic energy and decreasing with nucleon masses. According to the paper [6], for small number of nucleons, the negative energy corrections for nucleons can be compensated by positive relativistic corrections for potentials in the shell model.

The relativistic corrections in the external shells are significant and can achieve the value of spin–orbit interactions like for state $3p_{3/2}$ presented in Table 20.1. The results obtained show that the relativistic corrections in the shell model cannot be ignored.

Table 20.1.   The neutron's energy level E_{nlj}, spin-orbit interactions E_{ls} and relativistic corrections of masses ΔE_p in nucleus $_{79}\mathrm{Au}^{197}$.

nlj	E_{nlj}, MeV	E_{ls}, MeV	ΔE_p, MeV	$E_{nlj} + \Delta E_p$, MeV
$1s_{\frac{1}{2}}$	-45.55	—	-0.01323	-45.56
$1p_{\frac{3}{2}}$	-38.82	-0.4533	-0.02705	-38.85
$1p_{\frac{1}{2}}$	-37.46	$+0.9066$	-0.02705	-37.49
$1d_{\frac{5}{2}}$	-32.09	-0.9066	-0.09697	-32.19
$2s_{\frac{1}{2}}$	-31.19	—	-0.1159	-31.30
$1d_{\frac{3}{2}}$	-29.83	1.360	-0.09697	-29.93
$1f_{\frac{7}{2}}$	-25.35	-1.360	-0.1516	-25.51
$2p_{\frac{3}{2}}$	-24.44	-0.4533	-0.1929	-24.63
$2p_{\frac{1}{2}}$	-23.08	0.9066	-0.1929	-23.27
$1f_{\frac{5}{2}}$	-22.18	1.813	-0.1586	-22.34
$1g_{\frac{9}{2}}$	-18.62	-1.813	-0.2846	-18.90
$3s_{\frac{1}{2}}$	-16.81	—	-0.2717	-17.08
$1g_{\frac{7}{2}}$	-14.54	2.266	-0.2846	-14.82
$1h_{\frac{11}{2}}$	-11.89	-2.266	-0.1017	-11.99
$3p_{\frac{3}{2}}$	-10.08	-0.4533	-0.4131	-10.49
$3p_{\frac{1}{2}}$	-8.717	0.066	-0.4131	-9.130
$1h_{\frac{9}{2}}$	-6.907	2.720	-0.1017	-7.009

References

1. Janavičius, A. J. (1983). Relativistic corrections for one-neutron levels in square well of harmonic oscillator potential, *Inform. Higher Edu. Instit.*, **12**, 14–17 (in Russian).
2. Schiff, L. I. (1955). *Quantum Mechanics*, McGraw-Hill Book Company, Inc., New York, Toronto.
3. Jelley, N. A. (1990). *Fundamentals of Nuclear Physics*, Cambridge University Press, p. 278.
4. Solovjev, V. G. (1981). *Theory of Atomic Nucleus: Nuclei Models*, Moscow, Energoizdat, p. 296.
5. Shirokov, Yu. M. and Judin, N. P. (1972). *Nuclear Physics*, Moscow, Science, p. 671 (in Russian).
6. Veselov, A. I. and Kondratiuk, L. A. (1982). *Nuclear Physics*, **36**, p. 343.

Chapter 21

Relativistic Corrections to One-Nucleon Energy Levels for ^{208}Pb

21.1 Introduction

We have calculated the relativistic corrections to one-nucleon levels of ^{208}Pb in the first approximation of the perturbation theory for mass and potential energy in the Woods–Saxon potential. We have obtained corrections for the mass increase for the large principal or orbital quantum numbers. The corrections for the mass in the state $1i_{\frac{13}{2}}$ are obtained A[14] as -0.5663 MeV for neutrons and as -0.5884 MeV for protons. The corrections A[14] -0.6788 MeV for neutrons and -0.5884 MeV for protons for the excited states $1j_{\frac{15}{2}}$ and $1i_{\frac{13}{2}}$ are comparable with the energies of these levels. Corrections for the potential stay small for all states. Including the relativistic corrections for mass, we have obtained a better correlation between theoretical and experimental levels of energy for the bound and excited states of ^{208}Pb.

For calculating the energy levels of nuclei, the non-relativistic Hartree–Fock equations with the spin–orbit potential and effective interaction are used. According to [1], Vautherin and Brink introduced effective mass and Skyrme's forces that have improved this method. The same result was obtained from Dirac equation [2] taking into account that the binding energy E of one-particle states of nuclei is significantly less than the rest energy Mc^2. From that premise for the central potential case $U(r)$, the following equations were obtained:

$$\chi = \frac{1}{2M(r)}\vec{\sigma}\vec{p}\varphi, \tag{21.1}$$

$$E\varphi = \left(\vec{p}\frac{1}{2M(r)}\vec{p} + U(r) + \frac{1}{r}\frac{d}{dr}\left(\frac{1}{2M(r)}\right)\vec{l}\vec{\sigma}\right)\varphi, \tag{21.2}$$

for large φ and small χ components of the bispinor. Here, $M(r)$ is the effective mass. It was noted that [1] that inside a nucleus, $\frac{M(r)}{M} = 0.6$. One usually assumes that in the shell model $M(r) = M$, but this is not exact because mass of the nucleons depends on the state.

Binding energy is composed of only 1% of the rest energy. If we calculate 1% of the expectation value of the kinetic energy, we will obtain a significant value. The relativistic energy corrections for mass to the harmonic oscillator potential of the nucleus ^{197}Au A[14] reach $-0.4\,$MeV. In this chapter, the relativistic corrections for the Woods–Saxon potential were obtained for all bound neutron and proton states for the nucleus ^{208}Pb.

21.2   Semi-relativistic Equation

According to [2], the semi-relativistic Hamiltonian can be written in the form

$$\hat{H}_r = \hat{H}_m + \frac{\hat{p}^2}{2m} + \hat{H}_V + V(r) + V_{sl}(r). \tag{21.3}$$

The Hamiltonian includes relativistic corrections for mass and potential:

$$\hat{H}_m = -\frac{\hat{p}^4}{8m^3c^2}, \quad \hat{H}_V = -\frac{\hbar^2}{4m^2c^2}\frac{dV(r)}{dr}\frac{d}{dr}, \tag{21.4}$$

The term $V_{sl}(r)$ is the spin–orbital potential defining the arrangement of the energy levels of nucleons in the nuclei, i.e.,

$$V_{sl}(r) = -\kappa\frac{1}{r}\frac{dV(r)}{dr}(\vec{s}\vec{l}). \tag{21.5}$$

The magnitude of this potential κ is found from the distances between the energy levels of nucleons compared with the experiments. When the spin–orbit potential (21.5) is included, the spectroscopic notations $1s_{1/2}, 1p_{3/2}, 1p_{1/2}, 1d_{5/2}, 1d_{3/2}, 2s_{1/2}, \ldots$, of quantum states $nl_j (n = 1, 2, 3, \ldots, l = 0, 1, 2, 3, \ldots, l = s, p, d, f, \ldots,$ $j = l + 1/2, \ j = l - 1/2)$ and energy levels E_{nlj} are defined, where $n = 1, 2, 3, \ldots$ are the principal quantum numbers, orbital momentum $l = 0, 1, 2, 3$ and total momentum j of l, s quantum numbers are used. The levels of degeneration (the number of the same values of energies E_{nlj}) are $2j + 1$. In Ref. A[14], as in this chapter, the magnitude of the relativistic corrections for mass depends on the principal n and orbital l quantum numbers. These corrections were obtained for the same order as that for the spin–orbit interactions. Consequently, the determination of κ without taking into account the relativistic corrections for mass will be inexact. According to the operator identity $\hat{p}^2 = 2m\hat{T}$, where $\hat{T}$ is the kinetic energy operator from (21.3), we obtain

$$\hat{H}_m = -\frac{\hbar^4}{8m^3c^2} \left(\frac{d^4}{dr^4} + \frac{4}{r}\frac{d^3}{dr^3} - \frac{\hat{L}^2}{\hbar^2}\left(\frac{2}{r^2}\frac{d^2}{dr^2} + \frac{2}{r^4}\right) + \frac{\hat{L}^4}{\hbar^4}\frac{1}{r^2} \right).$$

$$(21.6)$$

We take the semi-relativistic equation for the Hamiltonian (21.3), and for the eigenfunction $R_\alpha = \frac{U_\alpha}{r}$ for the central forces case, we obtain the following equation:

$$c_1 \left(\frac{d^2 U_\alpha}{dr^4} + \left(\frac{1}{c_1} - \frac{2l(l+1)}{r^2} \right) \frac{d^2 U_\alpha}{dr^2} + \frac{4l(l+1)}{r^3}\frac{dU_\alpha}{dr} \right.$$
$$\left. + \frac{l^2(l+1)^2 - 6l(l+1)}{r^4} - \frac{ll+1}{c_1 r^2}U_\alpha \right)$$
$$+ c_2 r \frac{dV(r)}{dr}\frac{d}{dr}\frac{U_\alpha}{r} - c_0 V(r)U_\alpha - c_0 V_{sl}(r)U_\alpha = 0,$$

$$c_0 = \frac{2m}{\hbar^2}, \quad c_1 = \left(\frac{\hbar}{2mc}\right)^2, \quad c_2 = \frac{1}{2mc^2}. \qquad (21.7)$$

Substituting an asymptotic expression for the eigenfunction $U_\alpha \cong r^{\beta_\alpha}$ as $r \to 0$ in (21.7), we obtain

$$\beta_1 = l+1, \quad \beta_2 = -l, \quad \beta_3 = l+3, \quad \beta_4 = -l+2. \qquad (21.8)$$

Thus, we obtain the usual behavior $U_\alpha \cong r^{l+1}$ and the unusual one $U_\alpha \cong r^{l+3}$ at the origin. In this case, the relativistic corrections may be calculated in the first approximation as follows:

$$E_m = \int_0^\infty U_\alpha \hat{H}_m U_\alpha dr, \tag{21.9}$$

$$E_V = \int_0^\infty U_\alpha \hat{H}_V U_\alpha dr. \tag{21.10}$$

The matrix element for the relativistic corrections of mass may be expressed as follows [4]:

$$E_m = -\frac{\hbar^4}{8m^3c^2} \left(\int_0^\infty \left(\frac{d^2 U_{nlj}(r)}{dr^2} \right)^2 dr \right.$$

$$+ 2l(l+1) \int_0^\infty r^{-2} \left(\frac{dU_{nlj}(r)}{dr} \right)^2 dr$$

$$\left. + l(l+1)(l^2+l-6) \int_0^\infty r^{-4} U_{nlj}^2(r) dr \right). \tag{21.11}$$

21.3   Methods and Results

The Schrödinger equation may be solved by the discretization method [5]:

$$\frac{u_{i+1} - 2u_i + u_{i-1}}{h^2} + \rho(r_i)u_i = \lambda u_i, \tag{21.12}$$

$$\lambda = -bE, \quad i = 0, 1, 2, \ldots, n, \quad h = \frac{C}{n},$$

with the boundary conditions $u(0) = u(C) = 0$. We obtain non-relativistic eigenfunctions and eigenvalues solving the system of these equations (21.12) with the program EIGEN [5].

The program of numerical differentiation was formed according to the scheme of six points [6]

$$\left. \frac{d^k f}{dx^k} \right|_{x=\xi} = \sum_{i=0}^5 C_i f(x_i) + O(h^6 f^{(6)}), \quad 1 \le k \le 5. \tag{21.13}$$

The proposed method allows one to calculate the derivatives up to the fifth order.

The doubly magic ^{208}Pb nucleus is one of the best cases for studying the matrix elements for the residual two-body interactions in the nuclear shell model calculations and the simple schemes of one-nucleon energy levels in ^{209}Pb, ^{207}Pb, ^{207}Tl and ^{209}Bi [7].

The energy levels E_{nlj} of the neutrons and protons, expectation values of potential energies V_{nlj}, relativistic corrections for mass E_m and potential E_V of ^{208}Pb were calculated for the Woods–Saxon potential:

$$V_S(r) = -V_0^{N,Z}(1 + \exp\{\alpha^{N,Z}(r - R)\})^{-1} \qquad (21.14)$$

and spin–orbit potential (21.6) with parameters [7,8]

$$V_0^N = 44\,\text{MeV}, \quad V_0^Z = 60.3\,\text{MeV}, \quad \alpha^N = \alpha^Z = 1.5124\,\text{fm}^{-1},$$

$$R = 1.27A^{\frac{1}{3}}\,\text{fm}, \quad \kappa = 0.353\,\text{fm}^2.$$

The Coulomb potential was introduced in the usual form [7]:

$$V_C(r) = \frac{(Z-1)e^2}{4\pi\varepsilon_0 r} \begin{cases} \dfrac{3}{2}\dfrac{r}{R} - \dfrac{1}{2}\left(\dfrac{r}{R}\right)^3, & r \le R, \\ 1, & r > R. \end{cases} \qquad (21.15)$$

The relativistic corrections obtained for masses E_m and potential energies E_V for neutrons and protons are presented in Tables 21.1 and 21.2. Instead of V_{nlj}, the kinetic energy $E_k = E_{nlj} - V_{nlj}$ is presented. Bound energies E_{nlj} are compared with the experimental energies E_{ex}.

The calculated energy levels E_{nlj} for neutrons coincide with the experimental bound energy levels E_{ex} in Ref. [8]. We have obtained better coincidence in experiments for protons with $V_0^Z = 60.3\,\text{MeV}$ [8]. The inclusion of the relativistic corrections for the mass improves the results for all the excited states of the neutrons and the $3p_{\frac{3}{2}}$, $2f_{\frac{5}{2}}$, $3s_{\frac{1}{2}}$, $2d_{\frac{3}{2}}$, $1h_{\frac{11}{2}}$, $2d_{\frac{5}{2}}$, $1g_{\frac{7}{2}}$ states of protons for ^{208}Pb. The energy for the excited levels is the sum of the expectation values of the kinetic and potential energies. These energies can achieve magnitudes up to 40 MeV and have opposite signs. In that case, a

Table 21.1.   The neutrons' levels and relativistic corrections for ^{208}Pb.

nlj	E_{nlj}, MeV	E_k, MeV	E_m, MeV	$E_{nlj}+E_m$, MeV	E_{ex}, MeV	E_V, MeV
$3d_{\frac{3}{2}}$	-0.6957	21.01	-0.48340	-1.179	-1.34	-0.008
$2g_{\frac{7}{2}}$	-0.7670	25.62	-0.56160	-1.329	-1.39	0.004
$4s_{\frac{1}{2}}$	-1.4250	20.57	-0.39470	-1.820	-1.83	-0.010
$3d_{\frac{5}{2}}$	-1.9730	24.75	-0.53100	-2.504	-2.30	-0.007
$1j_{\frac{15}{2}}$	-1.9110	38.14	-0.67880	-2.681	-2.45	0.024
$1i_{\frac{11}{2}}$	-3.1590	30.46	-0.48420	-3.643	-3.09	0.032
$2g_{\frac{9}{2}}$	-3.8030	29.76	-0.56430	-4.367	-3.86	0.003
$3p_{\frac{1}{2}}$	-7.4710	24.18	-0.02950	-7.501	-7.36	0.004
$2f_{\frac{5}{2}}$	-8.1320	24.56	-0.22810	-8.360	-7.93	0.012
$3p_{\frac{3}{2}}$	-8.3700	25.18	-0.03600	-8.405	-8.25	0.003
$1i_{\frac{13}{2}}$	-8.6170	32.20	-0.56630	-9.183	-9.00	0.025
$2f_{\frac{7}{2}}$	-10.4200	26.27	-0.43890	-10.860	-10.30	0.010
$1h_{\frac{9}{2}}$	-10.9300	24.12	-0.41280	-11.340	-10.80	0.023
$1h_{\frac{11}{2}}$	-14.9300	25.58	-0.34690	-15.280	—	0.024
$3s_{\frac{1}{2}}$	-15.3800	21.66	-0.29430	-15.670	—	0.012
$2d_{\frac{3}{2}}$	-15.5600	20.94	-0.33670	-15.900	—	0.015
$2d_{\frac{5}{2}}$	-17.0100	21.68	-0.31290	-17.320	—	0.014
$1g_{\frac{7}{2}}$	-18.2100	19.37	-0.26530	-18.470	—	0.020
$1g_{\frac{9}{2}}$	-20.9400	20.17	-0.22560	-21.160	—	0.022
$2p_{\frac{1}{2}}$	-22.7000	16.48	0.00180	-22.700	—	0.015
$2p_{\frac{3}{2}}$	-23.4200	16.74	-0.00200	-23.420	—	0.015
$1f_{\frac{5}{2}}$	-24.8400	14.74	-0.15510	-24.990	—	0.017
$1f_{\frac{7}{2}}$	-26.5300	15.14	-0.13490	-26.660	—	0.019
$2s_{\frac{1}{2}}$	-29.5600	11.78	-0.08510	-29.650	—	0.014
$1d_{\frac{3}{2}}$	-30.7200	10.43	-0.08090	-30.800	—	0.013
$1d_{\frac{5}{2}}$	-31.6100	10.62	-0.07225	-31.680	—	0.015
$1p_{\frac{1}{2}}$	-35.7700	6.59	0.00250	-35.770	—	0.009
$1p_{\frac{3}{2}}$	-36.1300	6.66	0.00120	-36.130	—	0.010
$1s_{\frac{1}{2}}$	-39.9700	3.35	-0.00860	-39.980	—	0.005

small variation in the expectation values for the kinetic energies gives relatively significant exchange of the energy for the excited levels.

In our case, the insignificant relativistic corrections for potential energy of neutrons do not exceed 0.03 MeV, and for protons, they are less than 0.05 MeV for all states.

Table 21.2.   The protons' levels and relativistic corrections for ^{208}Pb.

nlj	E_{nlj}, MeV	E_k, MeV	E_m, MeV	$E_{nlj} + E_m$, MeV	E_{ex}, MeV	E_V, MeV
$3p_{\frac{3}{2}}$	-0.2865	28.37	-0.0595	-0.346	-0.62	0.017
$2f_{\frac{5}{2}}$	-0.0475	27.12	-0.5744	-0.622	-0.92	0.028
$1i_{\frac{13}{2}}$	-2.7190	34.64	-0.5884	-3.307	-2.15	0.050
$2f_{\frac{7}{2}}$	-3.2200	34.64	-0.5227	-3.743	-2.86	0.026
$1h_{\frac{9}{2}}$	-3.7810	25.91	-0.4940	-4.275	-3.76	0.040
$3s_{\frac{1}{2}}$	-7.8500	23.72	-0.3474	-8.202	-8.97	0.024
$2d_{\frac{3}{2}}$	-8.1620	22.69	-0.3965	-8.558	-9.32	0.029
$1h_{\frac{11}{2}}$	-9.3620	28.06	-0.4014	-9.763	-10.31	0.047
$2d_{\frac{5}{2}}$	-10.1500	23.65	-0.3618	-10.510	-10.64	0.028
$1g_{\frac{7}{2}}$	-11.6900	20.56	-0.3104	-12.000	-12.45	0.034
$1g_{\frac{9}{2}}$	-15.5500	21.90	-0.2967	-15.810	—	0.041
$2p_{\frac{1}{2}}$	-15.8300	17.62	0.0157	-15.840	—	0.026
$2p_{\frac{3}{2}}$	-16.8200	17.99	-0.1773	-16.830	—	0.027
$1f_{\frac{5}{2}}$	-18.7900	15.45	-0.1300	-18.970	—	0.028
$1f_{\frac{7}{2}}$	-21.2400	16.24	-0.0952	-21.380	—	0.024
$2s_{\frac{1}{2}}$	-23.1300	12.49	-0.08510	-23.220	—	0.021
$1d_{\frac{3}{2}}$	-24.9700	10.77	-0.0898	-25.060	—	0.021
$1d_{\frac{5}{2}}$	-26.3200	11.20	-0.0787	-26.400	—	0.025
$1p_{\frac{1}{2}}$	-30.1100	6.66	0.0045	-30.110	—	0.014
$1p_{\frac{3}{2}}$	-30.6900	6.85	0.0061	-30.700	—	0.017
$1s_{\frac{1}{2}}$	-34.1500	3.32	-0.0097	-34.160	—	0.008

21.4   Conclusions

The corrections for the mass increase when the expectation values of the kinetic energies of nucleons have maximum values. The biggest corrections for the mass of the neutrons in the state $1j_{\frac{15}{2}}$ are $-0.6788\,\mathrm{MeV}/c^2$ and those for the mass of the protons in the state $1i_{\frac{13}{2}}$ are $-0.5884\,\mathrm{MeV}/c^2$. These results are obtained for the greatest expectation values of kinetic energies 38.14 MeV and 34.64 MeV. However, there is no clear bond between these physical quantities. For example, the relativistic energy corrections for the mass of the neutrons for the upper filled state $3p_{\frac{1}{2}}$ are $-0.0295\,\mathrm{MeV}$ and those for state $1h_{\frac{9}{2}}$ are $-0.4128\,\text{MeV}$, but the expectation value of kinetic energy is almost the same, $-24.1\,\text{MeV}$. Consequently, the relativistic corrections for mass depend not only on the expectation value of kinetic energy but also on the potential. In the case of states $1s_{\frac{1}{2}}$, $1p_{\frac{3}{2}}$, $1p_{\frac{1}{2}}$, $1d_{\frac{5}{2}}$, $1d_{\frac{3}{2}}$ and $2s_{\frac{1}{2}}$, the magnitudes of corrections for the mass are insignificant for neutrons and protons.

The obtained results show that it is impossible to calculate the energies of the single-particle states without taking into account the relativistic corrections for mass. Hence, we must solve equation (21.8) if we want to obtain the exact values of energy levels.

This chapter used the program of high numerical accuracy for differentiation not only in the middle interval but also on the boundary. Without this program, it is impossible to calculate the relativistic corrections for mass (21.11) with sufficient accuracy.

References

1. Birbair, B. L., Savushkin, L. N. and Fomenko, V. N. (1982). *J. Nucl. Phys.*, **35**, 1134 (in Russian).
2. Schiff, L. I. (1955). *Quantum Mechanics*, McGraw-Hill, New York, Toronto, London, p. 473.
3. Karazija, R. (1987). *The Theory of X-Rays and Electronic Spectra of Free Atoms*, Mokslas, Vilnius, p. 274 (in Russian).
4. Ortega, J. M. and Pool, V. G. (1981). *An Introduction to Numerical Methods for Differential Equations*, Pitman Publishing Inc., New York, p. 288.
5. System/360 Scientific Subroutine Package. Programmer's Manual, N.Y., IBM Technical, p. 224.
6. Abramowitz, M. and Stegun, J. A. (eds.) (1964). *Handbook of Mathematical Functions*, National Bureau of Standards, New York, p. 832.
7. Valentin, L. (1982). *Subatomic Physic: Nuclei and Particles*, Vol. 2, Mir, Moscow, p. 330 (in Russian).
8. Soloviev, V. G. (1981). *Theory Atomic Nuclei. Nuclear Models*, Energoizdat, Moscow, p. 295 (in Russian).

Chapter 22

Solutions for the Semi-relativistic Equations for the Heaviest Nuclei

22.1 Introduction

The interest in the problem of the synthesis of superheavy atomic nuclei increased significantly within the past years. It is connected with successful Dubna experiments [1] on the synthesis of the $Z = 114$ isotopes with $A = 287, 288, 289$ where last nucleus with $N = 175$ neutrons has a last half-life of 30 s. Neutrons act as glue to hold protons together. Optimal relation N/Z corresponds approximately to 1.54. The announcement at Berkeley [2] was about detection of nuclei with $Z = 118$ in the ^{86}Kr + ^{208}Pb fusion reaction with large cross-section. Due to the strong Coulomb repulsion among protons, according to the liquid drop model, the nuclei would split immediately for $Z > 104$ at less values of N/Z. Only the quantum shell effects [3] allow very heavy elements to exist for a longer time. For consideration of stability of the external shells of the heaviest nuclei, the highly accurate mathematical methods must be used. In the external shells of superheavy elements, the rest interaction is small [4] and the shells' energies can be calculated using the Woods–Saxon single-particle potential [5] where relativistic corrections for the nucleons' mass or kinetic energy must be included A[13, 15].

The relativistic corrections for mass of nucleons in external shells, depending on state, can achieve -1.1 MeV, and significantly increase the binding energies of nucleons A[15] and increase stability of nuclei. In this case, we must solve the semi-relativistic differential equation of the fourth order with sufficient accuracy. For this aim, the semi-relativistic equation has been reduced to the integral–differential equation with the kernel, which is proportional to Green's function. It can be expressed by unperturbed wave functions and nonphysical solutions of the Schrödinger equation for the model potential. This method allows us to solve the semi-relativistic equation, where the relativistic corrections for mass and potential are included, with high accuracy A[15]. The corrections for the masses of nucleons are comparable with the energies of excited states and they are significantly increasing the stability of the external shells of heavy nuclei. The corrections for the potential are positive and small A[15], except for some light nuclei and do not play any role in the stability of nuclei. The energies of the one-nucleon levels E_{nlj} with the relativistic corrections for the mass E_m and the potential E_v for heavy nucleus [3] $^{285}_{114}X$, hypothetical nucleus [4] $^{310}_{126}X$ and proposed in A[16] $^{298}_{114}X$, $^{328}_{114}X$, $^{334}_{120}X$, $^{340}_{126}X$ like candidates to the more stable nuclei have been calculated for the spherically symmetric Woods–Saxon potential [5], A[16]

$$V_{\text{WS}}(r) = -V^{n,p}[1 + \exp[\alpha^{n \cdot p}(r - R)]]^{-1}, \qquad (22.1)$$

and the spin–orbit potential

$$V_{sl}(r) = -\kappa \frac{1}{r} \frac{d}{dr} V_{WS}(r)(\vec{\sigma} * \vec{l}), \qquad (22.2)$$

with the following parameters [5]:

$$\alpha^{n,p} = 1.5873\,\text{fm}^{-1}, \quad R = 1.24\,A^{1/3}, \quad V^{n,p} = V_m \left(1 \mp \gamma \frac{N-Z}{A}\right), \qquad (22.3)$$

$$\gamma = 0.63, \quad V_m = 53.3\,\text{MeV}, \quad \kappa = 0.263 \left(1 + 2\frac{N-Z}{A}\right)\,\text{fm}^2. \qquad (22.4)$$

These parameters were fitted to the one-nucleon levels [5, 8], A[15] in the region $15 \le A \le 340$. The Coulomb potential has been introduced in the usual form [6, 7]:

$$V_c(r) = \frac{(Z-1)e^2}{4\pi\varepsilon_0 r} P, \quad P = \frac{3r}{2R} - \frac{1}{2}\left(\frac{r}{R}\right)^3,$$

$$r \le R, \quad P = 1, \quad r > R. \tag{22.5}$$

22.2 The Integral–Differential Semi-relativistic Equation

If we consider relativistic corrections for mass and potential as perturbation, the semi-relativistic equation [7] can be expressed in the form

$$\frac{d^2}{dr^2}U_\alpha - \frac{l(l+1)}{r^2}U_\alpha + C[E_a - V_D - V_1(r)]U_\alpha = 0, \tag{22.6}$$

where we introduced the differential operator of the fourth order representing relativistic corrections $D(r)$

$$V_D = V_{\text{WS}}(r) + V_{sl}(r) - V_1(r) + \frac{C_1}{C}D(r)$$

$$+ C_1 r \left(\frac{d}{dr}V(r)\right)\frac{d}{dr}\frac{1}{r}, \tag{22.7}$$

$$C = \frac{2m}{\hbar^2}, \quad C_1 = \left(\frac{\hbar}{2mc}\right)^2, \quad L_0 = l(l+1),$$

$$D(r) = \frac{d^4}{dr^4} - \frac{2L_0}{r^2}\frac{d^2}{dr^2} + \frac{4L_0}{r^3}\frac{d}{dr} + \frac{(L_0)^2 - 6L_0}{r^4}. \tag{22.8}$$

Two last terms in (22.7) represent relativistic corrections for the nucleons' mass and potential energy correspondingly. The harmonic oscillator potential $V_1(r)$ A[9, 13]

$$V_1(r) = \frac{m\omega^2 r^2}{2}, \quad V_h(r) = V_1(r) + V_{\text{WS}}(0) \tag{22.9}$$

can express the model potential $V_h(r)$ and the average field $V_{\text{WS}}(r)$ of nucleus presented in Fig. 22.1.

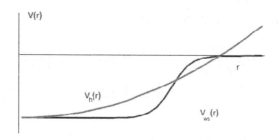

Fig. 22.1.   Comparing the harmonic oscillator $V_h(r)$ and Woods–Saxon V_{WS} potentials.

Introducing potentials $V_h(r)$ and $V_{WS}(r)$, we can get minimum of function $V_D(r)$ and maximum exact solutions of (22.6) for eigenfunctions $\Phi_2 U_{nl}$ (22.10) and eigenvalues E_α (22.9), (22.11) providing integration in the region $0 \leq r \leq r_{\max}$, $V_h(r_{\max}) = 0$. The radial wave functions A[9]

$$U_{nl} = e^{0.5\rho}\ \rho^{0.5(l+1)} \sum_{k=0}^{n-1} a_k \rho^k,$$

$$\rho = \frac{m\omega r^2}{\hbar}, \quad n = 1, 2, 3, \tag{22.10}$$

$$a_{k+1} = \frac{k - 0.5(\varepsilon_{nl} - l - 1.5)}{(k+1)(k+l+1.5)} a_k, \quad a_0 = 1, \tag{22.11}$$

and linearly independent nonphysical solutions A[9] for the model potential (22.9)

$$F_{nl} = e^{-0.5\rho} \rho^{-0.5l} \omega(\rho), \quad \omega = \sum_{k=0}^{\infty} b_k \rho^k, \tag{22.12}$$

$$b_{k+1} = \frac{k - 0.5(\varepsilon_{nl} + l - 0.5)}{(k+1)(k-l+0.5)} b_k, \quad b_0 = 1, \tag{22.13}$$

having the following eigenvalues:

$$E_{nl} = \varepsilon_{nl} \hbar\omega, \quad \varepsilon_{nl} = 2n + l - 0.5, \tag{22.14}$$

and Wronskian

$$W_0 = (2l + 1)\left(\frac{m\omega}{\hbar}\right)^{\frac{1}{2}}. \tag{22.15}$$

The eigenfunctions of Eq. (22.6), in the case of multiplicative perturbation theory A[9], must be expressed by multiplying the eigenfunctions U_{nl} for the model potential (22.11) $V_1(r)$ by the factor function A[15] which depends on the potential operator $V_D(r)$ for relativistic corrections, i.e.

$$U_\alpha = \Phi_{2,nlj} U_{nl}. \tag{22.16}$$

Substituting (22.16) into Eq. (22.6), we obtain this equation in the potential representation A[15]

$$U_{nl} = \frac{d^2}{dr^2} \Phi_2 + 2\left(\frac{d}{dr}\Phi_2\right)\left(\frac{d}{dr}U_{nl}\right) - CV_\delta U_{nl}\Phi_2 = 0, \tag{22.17}$$

$$V_\delta = V_D(r) - \Delta E_{nlj}, \qquad E_\alpha = E_{nl} + \Delta E_{nlj}. \tag{22.18}$$

Using the modified method of Lagrange A[15], the very handy integral equation was obtained

$$\Phi_2 U_{nl} = U_{nl} + \frac{U_{nl}}{W_0} \int_0^r F_{nl} C V_\delta \Phi_2 U_{nl} dr_1$$

$$- \frac{F_{nl}}{W_0} \int_0^r U_{nl} C V_\delta \Phi_2 U_{nl} dr_1, \tag{22.19}$$

$$\Delta E_{nlj} = \frac{\int_0^\infty U_{nl} V_D \Phi_{2,nlj} U_{nl} dr_1}{\int_0^\infty U_{nl} \Phi_{2,nlj} U_{nl} dr_1} \tag{22.20}$$

The obtained integral equation (22.19) was solved by the iteration method. For the zero approximation at the right-hand side of the integral equations, we must take $\Phi_2 = 1$ and then find ΔE_{nlj} from (22.20). We can freely choose the model potential (22.9), but it is better when unperturbed wave functions are close to perturbed wave functions $\Phi_2 U_{nl}$. Then a small number of the iterations provides highly accurate results. In our method, the frequency $\omega = d \cdot \omega_0$ for the model harmonic potential can be determined by the r.m.s. radius of the nuclei A[7]:

$$\omega_0 = 41 A^{-\frac{1}{3}} \frac{\text{MeV}}{\hbar}. \tag{22.21}$$

The constant d was found by variation in the interval $0.8 \le d \le 1.2$ demanding the minimum of the energy.

We verified the parameters for the Woods–Saxon potential (22.3), (22.4) calculating one-nucleon energies' levels of the doubly magic nucleus $^{208}_{82}$Pb taking in the care relativistic corrections for the mass and the potential (22.7). We used calculations, where relativistic corrections for mass are significant and coincide with experimental energy levels for some less interaction potential (22.3) for neutrons $0.98V^n$ and protons (22.3) V^p, which are presented in Table 22.1. The experimental meanings [6] of one-nucleon energy levels for protons E^p_{nlj} and neutrons E^n_{nlj} in Table 22.1 are presented like more black and represent good coincidence with the calculations. The doubly magic nucleus $^{208}_{82}$Pb is one of the best cases for shell-model calculations for definition A[15] and optimization of the Woods–Saxon potentials. Included relativistic corrections for masses E^p_m and E^n_m are significant and improve the accuracy of protons and neutrons bound and excited states of energies' levels with experimental results [8], A[14]. The obtained good coincidences in Table 22.1 propose a possibility for application of the Woods–Saxon potentials with some decreased $0.98V_m$ parameter (22.3) for neutrons of Chepurnov's potentials [5] considering stability of one $^{298}_{114}X$ of the heaviest nuclei A[16, 17]. The radial dependence of spin-orbit interaction (22.2) has maximum at the surface of nucleus and for excited states presented in Table 22.1 also must have strong dependence on significant relativistic corrections for the mass.

22.3   Results and Conclusions

At first, we calculated one-nucleon levels for the hypothetical nucleus $^{340}_{126}X$ and found closed proton–neutron shells with $Z = 114$, $N = 184$; $Z = 114$, $N = 214$; $Z = 120$, $N = 214$; $Z = 126$, $N = 214$ magic numbers of the pairs of protons and neutrons. It is interesting to investigate the stability of the nucleus $^{298}_{114}X$ with closed shells of protons and neutrons in a semi-relativistic approach using the potentials (22.1), (22.2), (22.5). We calculated one-nucleon energy levels of protons and neutrons of the nuclei $^{298}_{114}X$, $^{294}_{112}X$. The results are presented in Ref. A[17].

We used an expression of the kinetic energy for α particle $E_k = E_{A-4,Z-2} + E_\alpha - E_{A,Z}$ for the binding energies $E_{A,Z}$ of decaying

Table 22.1. The protons E^p_{nlj} and neutrons E^n_{nlj} of one-nucleon levels and relativistic corrections for mass E^p_m, E^n_m for nuclei $^{208}_{82}\mathrm{Pb}$.

Z	nlj	E^p_{nlj}, MeV	E^p_m, MeV	N	nlj	E^n_{nlj}, MeV	E^n_m, MeV
	$3p_{\frac{1}{2}}$	−0.198	−0.550		$3d_{\frac{3}{2}}$	−0.272	−0.490
	$3p_{\frac{3}{2}}$	−0.717	−1.02		$2g_{\frac{7}{2}}$	−0.310	−0.601
		−0.620					
	$2f_{\frac{5}{2}}$	−0.974	−0.803		$4s_{\frac{1}{2}}$	−1.21	−0.333
		−0.920					
	$1i_{\frac{13}{2}}$	−2.46	−0.587		$3d_{\frac{5}{2}}$	−1.72	−0.500
		−2.15				**−2.30**	
	$2f_{\frac{7}{2}}$	−3.24	−0.662		$1j_{\frac{15}{2}}$	−1.97	−0.846
		−2.86				**−2.45**	
	$1h_{\frac{9}{2}}$	−3.76	−0.381		$1i_{\frac{11}{2}}$	−2.72	−0.551
		−3.76				**−3.09**	
82	$3s_{\frac{1}{2}}$	−8.74	−0.387		$2g_{\frac{9}{2}}$	−3.73	−0.557
		−8.97				**−3.86**	
	$2d_{\frac{3}{2}}$	−9.36	−0.912	**126**	$3p_{\frac{1}{2}}$	−7.36	−0.485
		−9.32				**−7.36**	
	$1h_{\frac{11}{2}}$	−10.0	−0.368		$2f_{\frac{5}{2}}$	−7.94	−0.417
		−10.3				**−7.93**	
	$2d_{\frac{5}{2}}$	−10.8	−0.321		$3p_{\frac{3}{2}}$	−8.27	−0.480
		−10.6				**−8.25**	
	$1g_{\frac{7}{2}}$	−12.2	−0.281		$1i_{\frac{13}{2}}$	−8.74	−0.530
		−12.4				**−9.00**	
50	$1g_{\frac{9}{2}}$	−15.7	−0.234		$2f_{\frac{7}{2}}$	−10.4	−0.369
					$1h_{\frac{9}{2}}$	−10.8	−0.591
						−10.8	
				82	$3s_{\frac{1}{2}}$	−15.4	−0.315

Note: Z and N are numbers of protons and neutrons.

and daughter $E_{A-4,Z-2}$ nuclei and α particle $E_\alpha = 28.3\,\mathrm{MeV}$. Taking in the care that deeper one-particle levels presented in Table 22.2 for decaying $^{298}_{114}X$ and daughter $^{294}_{112}X$ nucleus practically coincide. We calculated $E_k = (742 + 28.3 - 783)\,\mathrm{MeV} = -12.7\,\mathrm{MeV}$ (disintegration energy $Q_k = -12.7\,\mathrm{MeV}$) and obtained that nucleus $^{298}_{114}X$ is stable A[17] with respect to the α-decay. The similar but less exact calculations without correction (22.1), (22.3) of potential $0.98V^n$ for the nucleus $^{298}_{114}X$ give stability A[16] but with less

Table 22.2.   The protons E^p_{nlj} and the neutrons E^n_{nlj} of one-nucleon levels and relativistic corrections for mass E^p_m, E^n_m for nucleus $^{298}_{114}X$.

Z	nlj	E^p_{nlj}, MeV	E^p_m, MeV	N	nlj	E^n_{nlj}, MeV	E^n_m, MeV
114	$1i_{\frac{13}{2}}$	−6.06	−0.457	**184**	$4s_{\frac{1}{2}}$	−6.30	−0.532
	$2f_{\frac{7}{2}}$	−6.52	−1.12		$3d_{\frac{3}{2}}$	−6.63	−0.951
	$1h_{\frac{9}{2}}$	−7.34	−0.300		$3d_{\frac{5}{2}}$	−7.06	−0.466
82	$3s_{\frac{1}{2}}$	−9.53	−0.240		$2g_{\frac{7}{2}}$	−7.28	−0.323
	$2d_{\frac{3}{2}}$	−10.4	−0.475		$1j_{\frac{15}{2}}$	−8.45	−0.681
	$1h_{\frac{11}{2}}$	−11.6	−0.259		$2g_{\frac{9}{2}}$	−9.79	−0.562
112	$1i_{\frac{13}{2}}$	−6.01	−0.392	**182**	$3d_{\frac{3}{2}}$	−5.41	−0.482
	$2f_{\frac{7}{2}}$	−6.14	−0.383		$3d_{\frac{5}{2}}$	−6.76	−0.470
	$1h_{\frac{9}{2}}$	−7.2	−0.230		$2g_{\frac{7}{2}}$	−6.87	−0.724
82	$3s_{\frac{1}{2}}$	−9.92	−0.296		$1j_{\frac{15}{2}}$	−8.10	−0.687
	$2d_{\frac{3}{2}}$	−10.6	−0.252		$2g_{\frac{9}{2}}$	−9.81	−0.450
	$1h_{\frac{11}{2}}$	−11.8	−0.250				

Note: Z and N are numbers of protons and neutrons.

disintegration energy A[17] $Q_k = -0.68\,\text{MeV}$. The obtained results coincide with the prediction of possibility of an island of relatively stable superheavy elements $^{298}_{114}X$, $^{328}_{114}X$, $^{334}_{120}X$, $^{340}_{126}X$ [8], A[16] with near magic neutrons $N = 184,\ 214$ and protons $Z = 114,\ 120,\ 126$ numbers. The reason of heavy nuclei stability is the balance between Coulomb forces and nuclear forces at double magi shells of protons and neutrons A[17]. In this case, heavy nuclei can have very long half-lives, reaching even the order of millions of years. From Table 22.1, we see that relativistic corrections for mass significantly increase the stability of nucleus $^{298}_{114}X$. The same calculations for nucleus $^{328}_{114}X$ approximately coincide with previous results A[16] (disintegration energy $Q_k = -0.701\,\text{MeV}$). From the presented shells of protons and neutrons in Ref. A[16] and Table 22.2 for the nucleus $^{298}_{114}X$, the β-decay is forbidden. Both nuclei $^{298}_{114}X$ and $^{328}_{114}X$ are stable according to β-decay. We obtained the nucleus $^{298}_{114}X$ at the following upper one-particle level $4s_{1/2}$ for neutrons $-6.30\,\text{MeV}$ and the second excited state $2f_{5/2}$ $-3.35\,\text{MeV}$ for protons. Then β-decay is forbidden for this case by the energy conservation law. Increasing ratio [9] N/Z increases stability of nuclei and optimum of stability corresponds to approximately 1.54. For nucleus $^{298}_{114}X$ we have relation 1.61. For proton state $1i_{13/2}$ of nuclei $^{298}_{114}X$ and $^{328}_{114}X$ we have decreasing Coulomb energies from 22.31 MeV to 21.81 MeV consequently A[16]. We have a similar situation for other proton states. Taking account of this fact and relativistic corrections to the mass of nucleons, we can suppose that nuclei $^{298}_{114}X$, $^{328}_{114}X$, $^{334}_{120}X$, $^{340}_{126}X$ can be stable A[16]. All results can be obtained only using the presented integral equations which can be solved with high accuracy for a mathematically complicated semi-relativistic task. The total nuclear energy evaluated by semi-empirical shell model calculations [11] does not coincide with stability calculations using the shell model. The important consequence from quantum diffusion theory of protons and neutrons in physical vacuum A[21] are the results of calculation of maximum radius of its spreading. We obtained $R_p = \hbar/\sqrt{2m|E|} = 10$ fm for protons in nucleus $^{208}_{82}$Pb using binding energy $E = -0.198$ MeV in the most excited state $3p_{1/2}$ and for state $3d_{3/2}$ of neutrons is obtained $R_n = 8.73$ fm taking E from Table 22.1. The same values of R are presented in Ref. A[19] considering the scattering of electrons and neutrons [13] on $^{208}_{82}$Pb nucleus. This result is the additional positive information that semi-relativistic calculations are exact even for excited states.

References

1. Oganessian, Yu. Ts. *et al.* (2000). The synthesis of superheavy nuclei in the ^{244}Pu $+ ^{48}$Cu reaction, *Nucl. Phys.* **63**(10), 1769–1779 (in Russian).
2. Ninov, V., Gregorich, K. E., *et al.* (1999). Nuclear energy: Chemical separation and purification of uranium, *Phys. Rev. Lett.* **83**, 1104–1110.
3. Smolanczuk, R. (1999). Production of superheavy elements, *Phys. Rev. C* **60**, 021301-1, 1–3.
4. Nazarevich, W. (1998). Frontiers of nuclear structure, *Nucl. Phys. A* **630**, 240–256.
5. Chepurnov, V. A. (1967). Average field of neutrons and protons. Shells with $N > 126$ and $Z > 82$, *J. Nucl. Phys.* **6**, 955–960 (in Russian).
6. Soloviev, V. G. (1982). *Theory of Atomic Nuclei. Nuclear Model*, Energoizdat, Moscow, p. 295 (in Russian).
7. Valentin, L. (1982). *Subatomic Physics: Nuclei and Particles 2*, Mir, Moscow, p. 330 (in Russian).
8. Jelley, N. A. (1990). *Fundamentals of Nuclear Physics*, Cambridge University Press, New York, p. 278.

9. Rydin, R. A. (2011). A new approach to finding magic numbers for heavy and super heavy elements, *Ann. Nucl. Energy* **38**, 238–242.

10. Moller, P. and Nix, J. R. (1976). Calculated half-lives of superheavy nuclei near ^{354}[126], *Phys. Rev. Lett.* **29**, 1461–1464.

11. Liran, S., Marinov, A. and Zeldes, N. (2000). Semiempirical shell model masses with magic number $Z = 126$ for superheavy elements. *Phys. Rev. C* **62**, 047301.

12. Strzalkowski, A. (1977). *Introduction to Physics of Nucleus*, National Publishing of Science, Poland, p. 637.

13. Yung-Kuo-Lim (2000). *Problems and Solutions on Atomic Nuclear and Particle Physics*, World Scientific, New Jersey, USA, p. 717.

Chapter 23

Stability of the Shells of the Heaviest Atomic Nuclei in the Semi-relativistic Model

23.1 Introduction

We have calculated one-nucleon energy levels for neutron and proton shells where the relativistic corrections to the mass and potential were included. We have found new candidates for the heaviest and more stable atomic nuclei $^{298}_{114}X$, $^{328}_{114}X$, $^{334}_{120}X$ and $^{340}_{126}X$.

The closed proton–neutron shells with $Z = 114$, $N = 184$; $Z = 114$, $N = 214$; $Z = 120$, $N = 214$; $Z = 126$, $N = 214$ magic pairs of proton and neutron numbers were found. The highly accurate method was used for obtaining the solution of the semi-relativistic equation. The corrections to the mass are comparable with the energies of protons and neutrons in the external shells and are important for consideration of the stability of the shells.

Due to the strong Coulomb repulsion between protons, according to the liquid drop model, the nuclei would split immediately for $Z > 104$. Only the quantum shell effects allow very heavy elements to exist. Recently, new heavy elements with $Z \geq 110$ were produced [1]. For consideration of stability of the external shells of the heaviest nuclei, the highly accurate mathematical methods must be used. In the external shells of superheavy elements, the rest interaction is small [2] and the shell energies can be calculated using the Woods–Saxon one-particle potential [3], where relativistic corrections to the nucleon mass or kinetic energy must be included A[15]. In this case, we must solve the semi-relativistic differential equation of the fourth order with sufficient accuracy. For this aim the semi-relativistic equation has been reduced to the integral–differential equation with the kernel proportional to Green's function. It can be expressed by unperturbed wave functions and nonphysical solutions of the Schrödinger equation for the model potential. This method allows us to solve the semi-relativistic equation, where the relativistic corrections to the mass and potential are included, with high accuracy A[15]. The corrections to the mass are comparable with the energies of excited states, and they increase the binding energies of the nucleons in the external shells of heavy nuclei. This is important for consideration of shell stability of heavy nuclei. The relativistic corrections to the potential are positive and approximately equal to the corrections of the mass A[15] and do not play any role in the shell stability of light nuclei. The energies of the one-nucleon levels E_{nlj} with the relativistic corrections to the mass E_m and the potential E_v of a heavy nucleus [3] $^{285}_{114}X$, hypothetical nucleus [2] $^{310}_{126}X$ and the nuclei $^{298}_{114}X$, $^{328}_{114}X$, $^{334}_{120}X$, $^{340}_{126}X$ were proposed in this chapter to be candidates for the more stable nuclei. We calculated them for the spherically symmetric Woods–Saxon potential [3], A[15]

$$V(r) = -V^{n,p}(1 + \exp\{\alpha^{n,p}(r - R)\}^{-1}), \tag{23.1}$$

and the spin–orbit potential

$$V_{sl}(r) = -\kappa \frac{1}{r}\frac{d}{dr}V(r)(\vec{\sigma} \cdot \vec{l}), \tag{23.2}$$

where we considered the following parameters:

$$\sigma^{n,p} = 1.5873\,\text{fm}^{-1}, \quad R = 1.24A^{1/3},$$

$$V^{n,p} = V_m\left(1 \mp \gamma\frac{N - Z}{A}\right), \tag{23.3}$$

$$\gamma = 0.63, \quad V_m = 53.3\,\text{MeV},$$

$$\kappa = 0.263\left(1 + 2\frac{N - Z}{A}\right)\text{fm}^2. \tag{23.4}$$

These parameters were fitted to the one-nucleon levels [3], A[15] in the region $15 \leq A \leq 209$. The Coulomb potential has been introduced in the usual form [4] as follows:

$$V_c(r) = \frac{(Z-1)e^2}{4\pi\varepsilon_0 r}P, \quad P = \frac{3r}{2R} - \frac{1}{2}\left(\frac{r}{R}\right)^3, \quad r \leq R,$$

$$P = 1, \ r\rangle R. \tag{23.5}$$

23.2 The Integral–Differential Semi-relativistic Equation

If we consider relativistic corrections to the mass and potential as perturbation, the semi-relativistic equation A[15] can be written in the form

$$\frac{d^2}{dx^2}U_\alpha - \frac{l(l+1)}{r^2}U_\alpha + C(E_\alpha - V_D - V_1(r))U_\alpha = 0, \tag{23.6}$$

where we introduced the differential operator of the fourth-order $D(r)$:

$$V_D(r) = V(r) + V_{sl}(r) - V_1(r) + \frac{C_1}{C}D(r) + C_1 r\left(\frac{d}{dr}V(r)\right)\frac{d}{dr}\frac{1}{r},$$

$$C = \frac{2m}{\hbar^2}, \quad C_1 = \left(\frac{\hbar}{2mC}\right)^2 \tag{23.7}$$

$$D(r) = \frac{d^4}{dr^4} - \frac{2L_0}{r^2}\frac{d^2}{dr^2} + \frac{4L_0}{r^3}\frac{d}{dr} + \frac{(L_0)^2 - 6L_0}{r^4}, \tag{23.8}$$

where $L_0 = l(l+1)$.

The two last terms in Eq. (23.7) represent the relativistic corrections to the nucleons' mass and potential energy (23.1), respectively. We used $V_1(r)$ as the model potential

$$V_1(r) = \frac{m\omega^2 r^2}{2}, \tag{23.9}$$

for the average field of the nucleus. The radial wave functions

$$U_{nl}(\rho) = e^{\frac{\rho}{2}} \rho^{\frac{l+1}{2}} \sum_{k=0}^{n-1} a_k \rho^k,$$

$$\rho = \frac{m\omega r^2}{\hbar}, \quad n = 1, 2, 3,$$
(23.10)

$$a_{k+1} = \frac{k - \frac{1}{2} \left(\varepsilon_{nl} - l - \frac{3}{2} \right)}{(k+1)\left(k + l + \frac{3}{2} \right)} a_k, \quad a_0 = 1,$$
(23.11)

and linearly independent non-physical solutions for the model potential (23.9)

$$F_{nl}(\rho) = e^{-\frac{\rho}{2}} \rho^{\frac{1}{2}} w(\rho), \quad w(\rho) = \sum_{k=0}^{\infty} b_k \rho^k,$$
(23.12)

$$b_{k+1} = \frac{k - \frac{1}{2} \left(\varepsilon_{nl} + l - \frac{1}{2} \right)}{(k+1)\left(k - l + \frac{1}{2} \right)} b_k, \quad b_0 = 1,$$
(23.13)

have the following eigenvalues:

$$E_{nl} = \varepsilon_{nl} \hbar \omega, \quad \varepsilon_{nl} = 2n + l - \frac{1}{2},$$
(23.14)

and the Wronskian

$$W_0 = (2l+1) \left(\frac{m\omega}{\hbar} \right)^{1/2}.$$
(23.15)

The eigenfunctions of Eq. (23.6) in the case of multiplicative perturbation theory A[9] must be expressed by multiplying the eigenfunction U_{nl} for the model potential $V_1(r)$ by the factor function A[15], which depends on the potential operator $V_D(r)$, i.e.,

$$U_\alpha = \Phi_{2,nlj} U_{nl}.$$
(23.16)

Substituting Eq. (23.16) into Eq. (23.6), we obtain this equation in the potential representation [7] as

$$U_{nl} \frac{d^2}{dr^2} \Phi_2 + 2 \left(\frac{d}{dr} \Phi_2 \right) \left(\frac{d}{dr} U_{nl} \right) - CV_\delta U_{nl} \Phi_2 = 0,$$
(23.17)

$$V_\delta(r) = V_D(r) - \Delta E_{nlj}, \quad E_\alpha = E_{nl} + \Delta E_{nlj}.$$
(23.18)

Using the modified Lagrange method A[15], a very handy integral equation can be obtained as follows:

$$\Phi_2 U_{nl} = U_{nl} + \frac{U_{nl}}{W_0} \int_0^r F_{nl} C V_\delta \Phi_2 U_{nl} dr_1$$

$$- \frac{F_{nl}}{W_0} \int_0^r U_{nl} C V_\delta \Phi_2 U_{nl} dr_1, \tag{23.19}$$

$$\Delta E_{nlj} = \frac{\int_0^\infty U_{nl} V_D \Phi_{2,nlj} U_{nl} dr_1}{\int_0^\infty U_{nl} \Phi_{2,nlj} U_{nl} dr_1}. \tag{23.20}$$

The integral equation (23.19) obtained was solved by the iteration method. For the zero approximation at the right-hand side of the integral equations, we must take $\Phi_2 = 1$ and then find ΔE_{nlj} from Eq. (23.20). We can freely choose the model potential (23.9), but it is better when unperturbed wave functions are close to perturbed wave functions $\Phi_2 U_{nl}$. Then a small number of the iterations provides highly accurate results. In our method, the frequency $\omega = d\omega_0$ for the model harmonic potential can be determined by the r.m.s. radius of the nuclei [5]

$$\omega_0 = 41 A^{1/3} \frac{\text{MeV}}{\hbar}. \tag{23.21}$$

The constant d was found by demanding the minimum of the energy.

23.3 Results and Conclusions

At first, we calculated the proton and neutron levels for the well-known nucleus $^{298}_{114}X$ and observed that not only shell effects but also relativistic effects for proton and neutron masses significantly increase its binding energies in heavy nuclei. For potentials (23.3)–(23.5), we have obtained the one-nucleon energy levels of protons in the states $3p_{3/2}$, $2f_{5/2}$ and $1i_{13/2}$ ($Z = 114$) is the number of protons when the ground state is filled) equal to -1.409 MeV, -2.103 MeV, and $-4,417$ MeV, respectively. For neutrons in the states $4s_{1/2}$, $3d_{3/2}$ and $2g_{7/2}$ ($N = 178$ is the number of neutrons when the ground state is filled), we obtained -7.326 MeV, -7.783 MeV and -9.049 MeV, respectively. Proton states $3p_{3/2}$, $2f_{5/2}$ and $1i_{13/2}$ ($Z = 114$) of the less stable isotope $^{285}_{114}X$ [1] with half-life time $T_\alpha = 800\,\mu s$ have the energies -0.6750 MeV, -1.464 MeV and -3.711 MeV. For neutrons in the states $4s_{1/2}$, $3d_{3/2}$ and $2g_{7/2}$ ($N = 178$) for $^{285}_{114}X$, we obtained -8.126 MeV, -8.244 MeV and -9.142 MeV. For the nucleus $^{298}_{114}X$ and the proton states $3p_{3/2}$, $2f_{5/2}$ and $1i_{13/2}$($Z = 114$), we obtained -2.816 MeV, -3.581 MeV and -5.939 MeV. For neutrons in the states $1k_{17/2}$, $2h_{11/2}$ and $4s_{1/2}$ ($N = 184$) we have -4.436 MeV, -5.715 MeV and -7.977 MeV. From the energies of external-shell protons $1i_{13/2}$ and neutrons $4s_{1/2}$, $3d_{3/2}$ and $2g_{7/2}$, we see that the nucleus $^{298}_{114}X$ is the most stable isotope as was supposed in Ref. [2]. Considering one-particle energies in closed shells and taking into account that after removing two protons from the shell $1i_{13/2}$, the energy of protons in this shell becomes -6.122 MeV, we obtained approximately the alpha-decay energy $Q_\alpha = 2.6921$ MeV, whereas for the well-known nucleus $^{239}_{94}$Pu [7], we obtained -5.25 MeV. From the presented proton and neutron shells' energies for the nucleus $^{298}_{114}X$, we see that beta-decay is forbidden. We have observed that the nucleus $^{298}_{114}X$ [1] is significantly more stable (calculated existence time 30 s) than $^{239}_{94}$Pu. We used one-particle approach, and our results are approximate. Using this approach for nuclei $^{288}_{114}X$ and $^{285}_{114}X$, we obtained alpha-decay energies $Q_{\alpha 1} = 8.384$ MeV and $Q_{\alpha 2} = 9.428$ MeV, respectively. These values can be compared with the experimental results $Q_{\alpha 1} = 9.85$ MeV [8] and $Q_{\alpha 2} = 11.18$ MeV [1].

After calculations, it has been found that the nucleus $^{340}_{126}X$ is more stable than $^{310}_{126}X$, which was supposed in Ref. [2] as a candidate for a more stable nucleus. For the closed shell of protons $3p_{1/2}$ ($Z = 126$) for the nucleus $^{340}_{126}X$, we obtained -3.227 MeV, but for $^{310}_{126}X$, this shell with parameter $V^p = 59.582$ MeV [3] was observed to be unbound. For $^{310}_{126}X$ with parameter $V^p = 59.582$ MeV [4], we obtained the following one-particle levels of protons: -0.22641 MeV, -0.7511 MeV, -1.110 MeV, -3.558 MeV for states $3p_{1/2}$ ($Z = 126$), $3p_{3/2}$, $2f_{5/2}$ and $1i_{13/2}$, respectively. In this case, the relativistic correction to the mass in the state $3p_{1/2}$ is -0.7973 MeV. For protons and neutrons, the negative corrections to the mass vary from -0.4 to -0.8 MeV and are significant when the nucleus stability is considered.

The relativistic corrections to the potential for the heavy nuclei are positive and they are below 0.04 MeV. The nucleus $^{340}_{126}X$ must have a large cross section of its formation because for $^{340}_{126}X$ we obtained the excited states $2g_{9/2}$, $1i_{11/2}$ and $1j_{15/2}$ protons with energies -1.103 MeV, -1.859 MeV and -2.164 MeV and the ground state $3p_{1/2}$ ($Z = 126$) energy -3.227 MeV. For one-particle neutron states $1l_{19/2}$, $1k_{17/2}$ ($N = 214$), $2h_{11/2}$, $4s_{1/2}$ and $3d_{3/2}$ for $^{340}_{126}X$, we found -0.5897 MeV, -6.623 MeV, -7.661 MeV, -9.581 MeV and -9.664 MeV, respectively. Undoubtedly, we have for states $1k_{17/2}$ and $4s_{1/2}$ the closed neutron shells with the appropriate numbers 214 and 184 of neutrons. For the neutron states $1l_{19/2}$, $1k_{17/2}$, $2h_{11/2}$, $4s_{1/2}$ ($N = 184$) and $3d_{3/2}$ of nucleus $^{310}_{126}X$, we obtained the similar energies -0.2303 MeV, -6.328 MeV, -7.628 MeV, -9.944 MeV and -10.05 MeV, respectively. We see that the energies of the one-particle states of neutrons of the different isotopes are close, but the energies of the external shells of protons for different isotopes differ significantly. In this situation, the magic number of protons is bound with some magic number of neutrons. We obtained interesting results for nuclei $^{304}_{120}X$ and $^{334}_{120}X$. For one-particle proton states $3p_{1/2}$, $3p_{3/2}$ and $2f_{5/2}$ ($Z = 120$) of $^{304}_{120}X$, we obtained energies -0.4550 MeV, -1.247 MeV and -2.119 MeV, and for $^{334}_{120}X$ we obtained, respectively, -4.700 MeV, -5.664 MeV and -6.501 MeV. Additionally, for $^{334}_{120}X$, we obtained the excited states of protons $2g_{9/2}$ and $1i_{11/2}$ with energies -2.993 MeV and -3.395 MeV. The energies of neutron states $4s_{1/2}$ ($N = 184$) of $^{304}_{120}X$ and $1k_{17/2}$ ($N = 214$) of $^{334}_{120}X$ are, respectively, -8.974 MeV and -5.870 MeV. From this and Ref. [2] we infer that many neutrons must be added in the valley of stability of the heaviest nuclei. But using stable nuclei in the fusion reactions, we can produce experimentally only proton-rich isotopes and it is impossible to reach the center of the island [9] $^{298}_{114}X$, $^{304}_{120}X$ and $^{334}_{120}X$, $^{310}_{126}X$, $^{340}_{126}X$ which are theoretically considered as nuclei by us. It is very important [9] for nuclear astrophysics investigations. Recent theoretical investigations [10] of fusion and fission cross-sections for the formation of a compound nucleus and survival probabilities allowed some experimental reactions for obtaining nuclei with $Z = 126$. Considering Table 23.1, where great bound energies -7.661 MeV for the 12 neutrons in subshell $2h_{11/2}$ are presented and paper's [10] predictions for perspective fusion reactions

$$^{66}_{28}\text{Ni} + ^{254}_{92}\text{Cf} \rightarrow ^{320}_{114}X + 2n, \quad ^{82}_{34}\text{Se} + ^{238}_{92}\text{U} \rightarrow ^{320}_{126}X, \quad (23.22)$$

we see their conforming with semi-relativistic model presented in this chapter and our papers A[16, 17].

Considering the plot of fission half-lives of even–even nuclei $^{252}_{100}\text{Fm}$, $^{254}_{102}\text{No}$ and $^{258}_{106}\text{Sg}$ [11], we see that for decreasing N/Z ratios for these isotopes with $N = 152$ and N/Z ratios 1.52, 1.49 and 1.43, the fission half-lives are decreasing correspondingly by 10^6 and 10^{12} times compared with the half-life $T_{1/2} \approx 10^{10}$s for $^{252}_{100}\text{Fm}$. We can find an approximate expression for the kinetic energy $E_k \approx Q_\alpha$ of an α particle through the binding energies $E_{A,Z}$ of decaying and daughter $E_{A-4,Z-2}$ nuclei $E_k = E_{A-4,Z-2} + E_\alpha - E_{A,Z}$ and the binding energy $E_\alpha = 28.3$ MeV of the α particle. Essentially, we used the bound energies of the nucleons in the external shells. Using the above-presented proton and neutron states for the nuclei $^{298}_{114}X$, $^{334}_{120}X$ and $^{340}_{126}X$ and energies -5.518 MeV, -6.453 MeV and -1.128 MeV of protons for states $1i_{13/2}$, $2f_{5/2}$ and $1j_{15/2}$ and energies -4.604 MeV, -5.440 MeV and -5.507 MeV of neutrons in states $2h_{11/2}$ and $1k_{17/2}$ for the nuclei $^{294}_{112}X$, $^{330}_{118}X$ and $^{336}_{124}X$, we obtained that for nuclei $^{298}_{114}X$, $^{334}_{120}X$ and $^{340}_{126}X$, we have $E_k \langle 0$. According to this more exact approach, we have found in the one-particle model that these nuclei with $E_k \langle 0$ are stable with respect to decay by α emission. The N/Z ratios for nuclei $^{298}_{114}X$, $^{334}_{120}X$ and $^{340}_{126}X$ arc, respectively, 1.61, 1.78 and 1.70, and they must have fission half-lives that are approximately 10^{12} or 10^{18} times greater. Results in Refs. [2, 10] can be explained by considering that neutrons do not repel each other by the Coulomb force but increase the volume of nuclei and decrease the repulsion forces between the protons.

Considering the experimental values of alpha-decay energies [1, 8], we can see that they are decreasing for more heavy isotopes. The one-nucleon levels E_{mlj}, magic numbers and relativistic corrections to the mass E_m^p of nucleons for hypothetical nucleus $^{340}_{126}X$ are presented in Table 23.1. From the above-considered heaviest atomic nuclei and Table 23.1, we obtained the following magic pairs $Z = 114$, $N = 184$; $Z = 114$, $N = 214$; $Z = 120$, $N = 214$; and $Z = 126$, $N = 214$

Table 23.1. One-nucleon levels of the protons E^p_{nlj} and the neutrons E^n_{nlj} and relativistic corrections to the mass E^p_m and E^n_m for a hypothetical nucleus $^{340}_{126}X$.

nlj	Z	E^p_{nlj}, MeV	E^p_m, MeV	nlj	N	E^n_{nlj}, MeV	E^n_m, MeV
$2g_{9/2}$	148	-1.103	-0.3975				
$1i_{11/2}$	138	-1.859	-0.3351	$1l_{19/2}$	234	-0.5879	-0.6352
$3p_{1/2}$	**126**	-3.227	-0.3837	$1k_{17/2}$	**214**	-6.623	-0.5149
$3p_{3/2}$	124	-4.414	-0.3837	$2h_{11/2}$	196	-7.661	-0.5746
$2f_{5/2}$	120	-4.716	-0.3430	$4s_{1/2}$	**184**	-9.581	-0.4829
$2i_{13/2}$	**114**	-7.507	-0.3259	$3d_{3/2}$	182	-9.664	-0.4963
$2f_{7/2}$	100	-7.553	-0.3381	$3d_{5/2}$	178	-10.75	-0.4513
$1h_{9/2}$	92	-8.828	-0.2687	$2g_{7/2}$	172	-10.90	-0.4978
$3s_{1/2}$	**82**	-10.31	-0.2345	$1j_{15/2}$	164	-12.20	-0.4497
$2d_{3/2}$	80	-11.17	-0.2045	$2g_{9/2}$	148	-13.40	-0.4034
$9h_{11/2}$	76	-12.17	-0.3114	$1i_{11/2}$	138	-13.85	-0.3254
$2d_{5/2}$	64	-12.71	-0.1925	$3p_{1/2}$	**126**	-15.95	-0.3562
$1g_{7/2}$	58	-13.81	-0.2057	$2p_{3/2}$	124	-16.57	-0.3156
$1g_{9/2}$	**50**	-15.40	-0.2093	$2f_{5/2}$	120	-17.07	-0.3165
$2p_{1/2}$	40	-17.36	-0.8244	$1i_{13/2}$	114	-17.49	-0.3193
$2p_{3/2}$	38	-17.79	-0.1213	$2f_{7/2}$	100	-18.93	-0.2938
$1f_{5/2}$	34	-19.45	-0.4987	$1h_{9/2}$	92	-19.81	-0.2623
$1f_{7/2}$	**28**	-21.15	-0.0857	$3s_{1/2}$	**82**	-22.46	-0.2039
$2s_{1/2}$	**20**	-22.33	-0.2261	$1h_{11/2}$	80	-22.47	-0.1969
$1d_{3/2}$	18	-23.95	-0.0530	$2d_{3/2}$	68	-23.20	-0.1998
$1d_{5/2}$	14	-24.67	-0.0975	$2d_{5/2}$	64	-24.06	-0.1715
$1p_{1/2}$	**8**	-27.23	-0.0308	$1g_{7/2}$	58	-24.35	-0.2043
$1p_{3/2}$	6	-27.97	-0.0472	$1g_{9/2}$	**50**	-25.28	-0.1985
$1s_{1/2}$	2	-29.83	-0.0094	$2p_{1/2}$	40	-28.73	-0.1732
				$2p_{3/2}$	38	-29.43	-0.1053
				$1f_{5/2}$	34	-29.74	-0.1078
				$1f_{7/2}$	**28**	-31.18	-0.0157
				$2s_{1/2}$	**20**	-33.94	-0.0517
				$1d_{3/2}$	18	-34.61	-0.0473
				$1d_{5/2}$	14	-34.99	-0.0453
				$1p_{1/2}$	**8**	-38.21	-0.0217
				$1p_{3/2}$	6	38.31	-0.0198
				$1s_{1/2}$	**2**	-41.35	-0.0108

Notes: Z and N are cumulative occupation numbers of protons and neutrons (the magic states and numbers are shown in bold).

of protons and neutrons for the most stable of them. Also, from Table 23.1 we can present a candidate to the stable heaviest nucleus $^{328}_{114}X$ with the magical number of neutrons $N = 214$ and with very high ratio 1.88 of neutrons to protons. The largest ratio $N/Z = 1.53$ was obatined in the reaction $^{48}_{20}Ca + ^{244}_{94}Pu = ^{289}_{114}X + 3n$ [9]. For one-particle states $2f_{5/2}$ and $1i_{13/2}$ ($Z = 114$) of $^{328}_{114}X$ for protons, we obtained energies -7.64 MeV and -10.3 MeV and for the state $1k_{17/2}$ ($N = 214$) of neutrons, we obtained the energy -4.88 MeV. Considering one-particle levels of nuclei $^{324}_{112}X$, $^{328}_{115}X$, and $^{328}_{113}X$, we obtain that the isotope $^{328}_{114}X$ is stable with respect to alpha- and beta-decays. The presented candidates for the stable isotopes $^{298}_{114}X$, $^{328}_{114}X$, $^{334}_{120}X$ and $^{340}_{126}X$ can be obtained only as a result of stellar nucleosynthesis because in the Earth's soil we do not have the stable nuclei with such a high ratio of neutrons to protons for producing super-heavy elements with the accelerators. Maybe super-heavy elements can be found in meteorites, in the Earth's mantle, or in the core.

References

1. Smolanczuk, R. (1999). Production of super-heavy elements, *Phys. Rev. C*, **60**, 021301-1, 1–3.
2. Nazarevich, W. (1998). Frontiers of nuclear structure, *Nucl. Phys. A*, **630**, 240–256.
3. Chepurnov, V. A. (1967). Average field of neutrons and protons. Shells with $N > 126$ and $Z > 82$, *J. Nucl. Phys.*, **6**, 955–960 (in Russian).
4. Soloviev, V. G. (1982). *Theory of Atomic Nuclei Nuclear Models*, Moscow, Energoizdat, p. 295 (in Russian).
5. Valentin, L. (1982). *Subatomic Physics: Nuclei and Particles 2*, Moscow, Mir, p. 330 (in Russian).
6. Jelley, N. A. (1990). *Fundamentals of Nuclear Physics*, Cambridge University Press, p. 278.
7. Enge, H. A. (1966). *Nuclear Physics*, Addison-Wesley, p. 582.
8. Oganessian, Yu. Ts. *et al.* (2000). The synthesis of superheavy nuclei in the $^{244}Pu + ^{48}Ca$ reaction, *Nucl. Phys.*, **63**(10), 1769–1777 (Russia).
9. Zagrebajev, V. and Greiner, W. (2010). New ideas on the production of heavy and superheavy neutron rich nuclei, *Nucl. Phys. A*, **834**, 366–369.
10. Manjunatha, H. C. and Sridhar, K. N. (2017). Projectile target combination to synthesis superheavy nuclei Z = 126, *Nucl. Phys. A*, **962**, 7–23.
11. Hofmanu, S. and Miinzenberg, G. (2000). The discovery of the heaviest elements, *Rev. Mod. Phys.*, **72**(3), 732–765.

Chapter 24

The Semi-relativistic Nuclear Shell Model for the Many-Particles Case

24.1 Introduction

In previous chapters, we have calculated the relativistic corrections for the mass and potential energy to one-nucleon levels, and significant terms of the relativistic corrections for the mass of nucleons were obtained. In this case, the mathematical problems of semi-relativistic model must be considered. The semi-relativistic equation is a differential equation of the fourth order, and it can be reduced to the integral–differential equation. The common solution of this equation must be expressed by superposition of four linear independent solutions.

Usually, we consider the nuclei like the non-relativistic systems. However, we must take into consideration [1] that the nuclear force has a repulsive core (-0.4 fm) and a great spin–orbit interaction. The repulsive core is generating the wave functions with high moments [1], and we cannot solve this problem A[15] using neither Schrödinger nor Hartree–Fock equations. The relativistic corrections for the mass or kinetic energy of nucleons must be included A[15] and the semi-relativistic Hamiltonian can be written in the following form:

$$H_r = H_m + \frac{\vec{p}^{\,2}}{2m} + H_\nu + V(r) + V_{sl}(r). \tag{24.1}$$

The first term of the Hamiltonian

$$H_m = -\frac{\vec{p}^4}{8m^3c^2},$$

(24.2)

and the third term

$$H_\nu = -\frac{\hbar^2}{4m^2c^2}\left(\frac{d}{dr}V(r)\right)\frac{d}{dr},$$

(24.3)

include the relativistic corrections to the mass of nucleons and the potentials [2] of their interaction. The following term:

$$V_{sl}(r) = -\kappa\frac{1}{r}\left(\frac{d}{dr}V(r)\right)(\vec{\sigma}*\vec{l}).$$

(24.4)

is the spin–orbital potential, which also has the relativistic origin. The semi-relativistic equation for the eigenfunction

$$R_\alpha = \frac{U_\alpha}{r},$$

can be obtained A[15] from the Hamiltonian (24.1) for the central potential $V(r)$, spin–orbit potential $V_{sl}(r)$ and model potential $V_1(r)$ in the following form:

$$\frac{d^2}{dr^2}U_\alpha - \frac{L(L+1)}{r^2}U_\alpha + C(E_0 - V_D - V_1(r))U_\alpha = 0,$$

(24.5)

where we introduced the differential operator $D(r)$ of the fourth order

$$V_D = V(r) + V_{sl}(r) - V_1(r) - \frac{C_1}{C}D(r)$$

$$- C_1r\left(\frac{d}{dr}V(r)\right)\frac{d}{dr}\frac{1}{r},$$

(24.6)

$$C = \frac{2m}{\hbar^2}, \quad C_1 = \left(\frac{\hbar}{2mc}\right)^2, \quad CC_1 = \frac{1}{2mc^2}$$

$$D(r) = \frac{d^4}{dr^4} - \frac{2L_0}{r^2}\frac{d^2}{dr^2} + \frac{4L_0}{r^3}\frac{d}{dr} + \frac{(L_0)^2 - 6L_0}{r^4},$$

(24.7)

$$L_0 = L(L+1).$$

The forth and the fifth terms in the (24.6) represent the relativistic corrections for the mass of nucleons and potential. Substituting the asymptotic expression of the eigenfunction $U_\alpha \cong r^\beta$ as $r \to 0$ in (24.5), we obtain four partially independent asymptotic solutions

$$U_{\alpha 0} \cong r^{L+1}, \quad F_{\alpha 0} \cong r^{-L}, \quad U_{\alpha 1} \cong r^{L+3}, \quad F_{\alpha 1} \cong r^{-L+2}. \quad (24.8)$$

Then, the general solution of the linear differential equation of the fourth order can be expressed as a linear combination of these linear independent solutions, i.e.,

$$U_\alpha(r) = C_{\alpha 0} U_{\alpha 0}(r) + C_{0\alpha} F_{\alpha 0}(r) + C_{\alpha 1} U_{\alpha 1}(r) + C_{1\alpha} F_{\alpha 1}(r). \quad (24.9)$$

The physical solution $U_{\alpha 0} \cong r^{L+1}$ and non-physical solution $F_{\alpha 0} \cong r^{-L}$ show different behaviors at the origin for $r \to 0$ and are named as physical and non-physical solutions A[15] of the radial Schrödinger equation. For the semi-relativistic equation (24.5), we obtained the complementary physical solution $U_{\alpha 1} \cong r^{L+3}$. For orbital angular momentum quantum numbers $L = 0$ and $L = 1$, we may have a regular physical solution $F_{\alpha 1} \cong r^{-L+2}$ at the origin. Assuming that the potential energies $V(r)$ and $V_{sl}(r)$ vanish at great distances, we can find the asymptotic expression of the differential equation (24.5) as follows:

$$C_1 \frac{d^4}{dr^4} U_\alpha(r) + \frac{d^2}{dr^2} U_\alpha(r) + C E_\alpha U_\alpha(r) = 0, \quad (24.10)$$

and the following four asymptotic solutions of (24.5) in the exponential form $U_\alpha \cong \exp(k_\alpha r)$. In this case, we obtain

$$k_{\alpha 1} = -k_\alpha, \quad k_{\alpha 2} = k_\alpha,$$

$$k_\alpha = \frac{1}{\sqrt{2C_1}} (-1 + \sqrt{1 - 4C_1 C E_\alpha})^{1/2},$$

$$k_{\alpha 3} = i k_{\alpha m}, \quad k_{\alpha 4} = -i k_{\alpha m}, \quad (24.11)$$

$$k_{\alpha m} = \frac{1}{\sqrt{2C_1}} (1 + \sqrt{1 - 4C_1 C E_\alpha})^{1/2}.$$

Usually, $4C_1 C E_\alpha \langle 1$, and we can use the approximate expression $k_\alpha \approx \sqrt{-C E_\alpha}$. The physical solutions of (24.5) can be expressed in the following form:

$$U_\alpha(r) = r^\beta W(r) \exp\{-k_\alpha r\}. \quad (24.12)$$

24.2 The Solutions of Integral–Differential Semi-relativistic Equation for the Singular Potentials

The integral–differential semi-relativistic equation (24.5) was considered in Ref. A[15]. The semi-relativistic solutions represent more tightly bounded nucleons and are decreasing at infinity faster than the solutions of the Schrödinger equation. This fact is important when we consider the stability of the shells of a nucleon for a superheavy nucleus. However, in the origin, we have two kinds of different physical solutions (24.8) $U_{\alpha 0}$ and $U_{\alpha 1}$. The semi-relativistic solutions $U_{\alpha 0}$ can be used for non-singular or for singular potentials $V_{\alpha 0}(r)$, which behave at the origin like $r^{-\gamma}$ where $0\langle\gamma\langle 4$. The first kind of solutions $U_{\alpha 0}$ of Eq. (24.5) can be used for the Coulomb, Yukawa and spin–orbit potentials. These solutions have the asymptotes at the origin r^{L+1} that are similar to the wave functions of the Schrödinger equation. In the non-relativistic approach [3], for this case, we can find the nonphysical solutions $F_{\alpha 0}$ with the behavior at the origin similar to r^{-L}. In addition, we can use the second kind of semi-relativistic solutions $U_{\alpha 1}$ where singular spin–orbit potentials [1] with the singularity r^{-3} can be included, since the semi-relativistic equation (24.5) has the asymptotic r^{-4} in the origin. Using solutions $U_{\alpha 1}$, we can find the expectation values for all terms of the realistic Hamada–Johnston potential for the OPEP approach. Using the solutions $U_{\alpha 1}$ for the Hamada–Johnston potential with singularity r^{-6}, all integrals of the expectation values can be calculated.

For convenience, we introduce a dimensionless parameter $\rho = \frac{r}{F}$ and then the radial semi-relativistic equations (24.5)–(24.7) can be presented in the following form:

$$C_F D(\rho) U_\alpha + C_F C_\rho \left(\frac{d}{d\rho} V \right) \frac{d}{d\rho} \frac{U_\alpha}{\rho} + \frac{d^2}{d\rho^2} U_\alpha$$

$$- \frac{L(L+1)}{\rho^2} U_\alpha + CF^2 (E_\alpha - V) U_\alpha = 0,$$

$$C_F = \frac{C_1}{F^2}, \quad C_2 = C_1 C. \tag{24.13}$$

Now we can see that for large F the semi-relativistic equation reduces to the Schrödinger equation. For the nucleons localized around the center of force about $1\,\text{fm}$, we have $C_F = 0.011$, in the region of a repulsive core $(0.4\,\text{fm})$, we have $C_F = 0.07$. For the electrons in the first Bohr orbit, we obtain $C_F = 1.3 \cdot 10^{-15}$. These results show that in the theory of electronic spectra we can calculate the relativistic corrections with sufficient accuracy in the first approximation of the perturbation theory A[15]. However, for calculations of the nuclear energy levels, we must include higher order perturbations as well as solutions $U_{\alpha 0}$ and $U_{\alpha 1}$ of the semi-relativistic equation (24.5).

24.3 The System of Integral Semi-relativistic Equations in the Hartree–Fock Approach

A general solution of the Schrödinger equation in the potential representation can be presented in the form of integral equations A[9]. In this representation, the wave functions are expressed as a product of the unperturbed solution and the function which depends on the perturbation potential and relativistic corrections. This method was used A[15] for the one-particle case for calculations of relativistic corrections in the average field of neutron and proton shells, and now we can use this method for a many-particle case in the simple Hartree–Fock approach. Suppose that nucleons fill up a number of nucleon orbitals so that they form a wave function in terms of occupied one-particle states:

$$U_{\alpha p} = U_{\alpha p}(\xi_1, \xi_2, \dots, \xi_n), \qquad (24.14)$$

and the multiplier Φ depending on two-body interaction potentials [1] $V(\vec{r}, \vec{r}')$. Here, we included the functions, where ξ can be a space, a spin or isospin coordinates. U_α are anti-symmetrical combinations of wave functions $f(\xi_i)U_{\alpha 1}(\xi_i)$ expressed by multiplication of one-nucleon wave functions $U_{\alpha 1}(\xi_i)$ for the Woods–Saxon potential, where f are the functions which depend on two-body interaction potential. Using the Jastrow correlation method [3], the wave function for A nucleons can be expressed in the following way:

$$\Psi(\vec{r}_1, \vec{r}_2, \ldots, \vec{r}_A) = (C_A)^{-\frac{1}{2}} \prod_{1 \leq i \langle j \leq A} f(|\vec{r}_i - \vec{r}_j|) \Phi(\vec{r}_1, \vec{r}_2, \ldots, \vec{r}_A),$$

(24.15)

where C_A is the normalization constant, Φ is a single Slater determinant, and f denotes an independent correlation factor.

The Hartree–Fock equations can be written in a short form as [1]

$$-\frac{\hbar^2}{2m} \Delta \Psi_i(\vec{r}) + V_D(\vec{r}) \Psi_i(\vec{r}) + V_H(\vec{r}) \Psi_i(\vec{r})$$

$$- \int V_F(\vec{r}, \vec{r}') \Psi_i(\vec{r}') \vec{dr}^l = \varepsilon_i \Psi_i(\vec{r}),$$

(24.16)

where the Hartree–Fock terms including exchange effects are presented as follows:

$$V_H(\vec{r}) = \sum_{\alpha \in F} \int \Psi_\alpha^*(\vec{r}') V_F(\vec{r}, \vec{r}') \Psi_\alpha(\vec{r}') \vec{dr}^l,$$

(24.17)

$$V_F(\vec{r}, \vec{r}') = \sum_{\alpha \in F} \int \Psi_\alpha^*(\vec{r}') V(\vec{r}, \vec{r}') \Psi_\alpha(\vec{r}) \vec{dr}.$$

(24.18)

For the spherically symmetric potentials (24.6) $V_D(r)$, where relativistic corrections can be included A[15], in the case of multiplicative perturbation theory A[15], the eigenfunctions U_α must be expressed by multiplying the eigenfunctions A[9] $U_{nL}(r_i)$ for the model potential $V_1(r) = \frac{m\omega^2 r^2}{2}$ by the factor function Φ_{nLj}, which depends on $V_D(r)$, i.e,

$$U_\alpha = \Phi_{2, nLj} U_{nL}.$$

(24.19)

Then, applying the potential representation method A[9] to the Hartree–Fock equation, we can obtain

$$U_{nL} \frac{d^2}{d\rho^2} \Phi_2 + 2\left(\frac{d}{dr}\Phi_2\right)\left(\frac{d}{dr}U_{nL}\right) - CV_\delta \Phi_2 U_{nL} = 0,$$

(24.20)

$$V_\delta(r) = V_D(r) + V_H(\vec{r})U_\varepsilon(\vec{r})$$

$$+ C \int V_F(\vec{r}, \vec{r}')U_\alpha(\vec{r}')d\vec{r}'^{\,l} - \Delta E_{nLj}, \qquad (24.21)$$

$$V_F(\vec{r}, \vec{r}') = \sum_{\alpha \in F} U_\alpha^* V(\vec{r}, \vec{r}')U_\alpha(\vec{r}) \qquad (24.22)$$

$$V_\delta(r) = \Delta E_{nLj}, \quad E_\alpha = E_{nL} + \Delta E_{nLj}. \qquad (24.23)$$

Using the modified method of Lagrange A[15], the very handy system of integral equations can be obtained

$$\Phi_2 U_{nL} = U_{nL} + \frac{U_{nL}}{W_0} \int_0^r F_{nL} C V_\delta \Phi_2 U_{nL} dr_1$$

$$- \frac{F_{nL}}{W_0} \int_0^r U_{nL} C V_\delta \Phi_2 U_{nL} dr_1, \qquad (24.24)$$

$$\Delta E_{nLj} = \frac{\int_0^\infty U_{nL} \Delta V_D \Phi_2 U_{nL} dr_1}{\int_0^\infty U_{nL} \Phi_{2,nLj} U_{nL} dr_1} \qquad (24.25)$$

Here, U_{nL} and F_{nL} are physical and nonphysical solutions A[9] for the model harmonic oscillator potential. For these linearly independent solutions, we can obtain their Wronskian

$$W_0 = (2L + 1) \left(\frac{m\omega}{\hbar} \right)^{1/2}. \qquad (24.26)$$

The system of integral equations (24.24) and (24.25) can be solved by the iteration method. If we want to find the physical solutions with the asymptote in the origin r^{L+3}, we must use the solutions U_{nL+2} and F_{nL+2} and find a new expression for W_0. In this case, we must compensate the change from L to $L+2$ in physical and nonphysical solutions and in W_0. Then we must take the expression

$$V_{Dl}(r) = V_d(r) + \frac{1}{C} \frac{4L + 6}{r^2} \qquad (24.27)$$

instead of $V_D(r)$ in (24.21).

In this case, using (24.24) and (24.25), we can obtain a set of physical solutions that show a more rapid decrease in the origin and compatible with realistic Hamada–Johnston potentials.

24.4   Conclusions

In this chapter, we presented two different sets of physical solutions with different asymptotes at the origin. The simple Hartree–Fock approach is presented for finding these solutions using the multiplicative or potential representation method of the perturbation theory. The important relativistic corrections for mass of nucleons are included. The presented semi-relativistic approach can be used in estimating the energy levels of a heavy-nucleus and in estimating the stability of proton and neutron shells in superheavy nuclei.

References

1. Heyde, K. L. G. (1994). *The Nuclear Shell Model*, Springer-Verlag, Berlin.
2. Shiff, L. I. (1955). *Quantum Mechanics*, McGraw-Hill Book Company, New York, p. 457.
3. Stoitsov, M. V., Antonov, A. N. and Dimitrova, S. S. (1993). Natural orbital representation and short-range correlations in nuclei, *Phys. Rev. C* **48**, 75.

Chapter 25

Relativistic Corrections for Different States of the Charmed and Bottom Mesons

25.1 Introduction

In this chapter, the mathematical methods for the semi-relativistic model, presented for the nuclei in earlier chapters, are applied to charmed and bottom mesons. The semi-relativistic single-particle equation is a differential equation of the fourth order, and it can be reduced to the integral–differential equations. The general solution of this equation must be expressed by the superposition of the four linearly independent solutions. Developing the modified method of Lagrange and the multiplicative perturbation theory we obtained the integral–differential equations for the wave functions with the usual asymptote at the origin r^{L+1} and unusual asymptote at r^{L+3} and r^{-L+2}. For ground state of hadrons $n = 1$, $L = 0$, we obtained three different eigenfunctions with different eigenvalues. The wave functions with the asymptote at the origin r^{L+3} must be used when the singular realistic nuclear potentials are included. We described the spectrum of the bottomonium and charmonium states sufficiently exactly using the semi-relativistic model for the harmonic oscillator potential.

Often, we consider the nuclei and quarks as the non-relativistic systems. However, we must take into consideration [1] that the nuclear force has a repulsive core $(-0.4\,\text{fm})$ and a great spin–orbit interaction, which has a relativistic descent. We cannot solve the task of the energy spectra for a nucleus without including the relativistic corrections to the mass A[15] and the repulsive core generated by the wave functions with high momentum [1]. The interaction potential, even for heavy quarks, for any calculations of excited states' energies is not based on realistic physical and mathematical assumptions. The aim of this chapter is to obtain the integral equations for solutions of semi-relativistic equation where these corrections are included.

We must begin from the consideration of the semi-relativistic Hamiltonian for one-particle or relative two-particle movement with reduced mass in the COM (center-of-mass) frame with the following additional terms in the Hamiltonian A[15]:

$$H_m = -\frac{\vec{p}^{\,4}}{8m^3c^2}, \tag{25.1}$$

$$H_\nu = \frac{\hbar^2}{4m^2c^2}\left(\frac{d}{dr}V(r)\right)\frac{d}{dr}, \tag{25.2}$$

which represents the relativistic corrections to the mass for nucleons or quarks and the interaction potential. For the Hamiltonian, with the additional terms (25.1) and (25.2), we obtained the fourth-order differential semi-relativistic equation for the central potential $V(r)$ in the special form [2]:

$$C_F D(\rho)U_\alpha + C_F C_\rho\left(\frac{d}{d\rho}V\right)\frac{d}{d\rho}\frac{U_\alpha}{\rho} + \frac{d^2}{d\rho^2}U_\alpha$$

$$- \frac{L_0}{\rho^2}U_\alpha + CF^2(E_\alpha - V)U_\alpha = 0, \tag{25.3}$$

$$D(\rho) = \frac{d^4}{d\rho^4} - \frac{2L_0}{\rho^2}\frac{d^2}{d\rho^2} + \frac{4L_0}{\rho^3}\frac{d}{d\rho} + \frac{(L_0)^2 - 6L_0}{\rho^4}, \quad L_0 = L(L+1), \tag{25.4}$$

$$C = \frac{2m}{\hbar^2}, \quad C_1 = \left(\frac{\hbar}{2mc}\right)^2, \quad C_F = \frac{C_1}{F^2}, \quad r = \rho \cdot F. \tag{25.5}$$

Here, we introduced a dimensionless parameter $\rho = r/F$ using scale of distances expressed in Fermi F.

The first term and the second term in (25.3) represent the relativistic corrections for the mass of nucleons or quarks and for the potential $V(r)$. Substituting the asymptotic expression of the eigenfunctions $U_\beta \cong \rho^\beta$ as $\rho \to 0$ in (25.6), we obtained the following four partial independent asymptotic solutions A[19, 20]

$$U_{\beta 1} = \rho^{L+1}, \quad U_{\beta 2} = \rho^{-L}, \quad U_{\beta 3} = \rho^{L+3}, \quad U_{\beta 4} = \rho^{-L+2}. \quad (25.6)$$

Then, a general solution of the fourth-order linear differential equation can be expressed as a linear combination of these linear independent solutions A[19].

The wave functions $U_{\alpha i} = \varphi_{\alpha i} U_{\beta \cdot i}$ with asymptotes $U_{\alpha 1} \cong \rho^{L+1}$ and $U_{\alpha 3} \cong \rho^{L+3}$ and singular solutions $U_{\alpha 2} \cong \rho^{-L}$ show a different behavior at the origin for $\rho \to 0$ and are named as relativistic equation (25.3). We obtained the complementary physical solutions $U_{\alpha 3} \cong \rho^{L+3}$ for $L = 0, 1, 2, \ldots$. For quantum numbers $L = 0$ and $L = 1$ of the orbital angular momentum, we have two additional semi-relativistic regular physical solutions $U_{\alpha 4} \cong \rho^{-L+2}$ and $U_{\alpha 3} \cong \rho^{L+3}$ at the origin.

Assuming that the potential energies $V(r)$ vanish at great distances, we found A[15] for this case asymptotic solutions of (25.3) at large distances in the exponential form $U_\alpha \cong e^{k_\alpha r}$, $k_\alpha \approx -\sqrt{-CE_\alpha}$. We obtained that the solutions for the semi-relativistic k_α represented the more tightly bounded nucleons or quarks. Considering the asymptotic solutions (25.6), we can use for bound states the following approximate expression:

$$U_{\alpha i} = \varphi_{\alpha i} \rho^{\beta(i)}, \quad i = 1, 2, 3, 4. \quad (25.7)$$

25.2 The Solutions of the Integral–Differential Semi-relativistic Equation

The semi-relativistic equation (25.3) was considered in Ref. A[15] like some perturbation for the Schrödinger equation where the solutions $U_{\alpha 3}$ and $U_{\alpha 4}$ were not included. The semi-relativistic solutions in this case represent more tightly bounded systems and are decreasing at infinity faster than the solutions of the Schrödinger equation. This fact is important in consideration of one-nucleon levels A[15] and the stability of the nucleon shells for the superheavy nuclei A[16]. For the semi-relativistic equation at the origin, we have three kinds of different physical solutions $U_{\alpha 1}$, $U_{\alpha 3}$ and $U_{\alpha 4}$ for quantum numbers $L = 0$ and $L = 1$. For the same quantum numbers and for $L > 1$, we have two physical semi-relativistic solutions $U_{\alpha 1}$ and $U_{\alpha 3}$. For the Schrödinger equation, where relativistic corrections are included and the standard perturbation theory is used, we have only one physical solution $U_{\alpha 1}$.

The semi-relativistic solutions $U_{\alpha 1}$ can be used for non-singular or for singular potentials which behave at the origin like $\rho^{-\gamma}$, where $0 < \gamma < 2$. The solutions $U_{\alpha 1}$ can be used for Coulomb, Yukawa, Woods–Saxon and spin–orbit potentials [1] for the average field of nucleus and quark–antiquark interaction at short distances [2]. These potentials have the asymptote ρ^{-1} at the origin, and semi-relativistic solutions with asymptote ρ^{L+1} are similar to the wave functions of the Schrödinger equation. Also, in the semi-relativistic approach A[15], like in the Schrödinger equation's case, we have the non-physical solutions $U_{\alpha 2}$ with the behavior at the origin similar to ρ^{-L}. In the following formulas, we will mark the solutions $U_{\alpha 2}$ as F_{nL} as was done in the paper A[15]. We can use the second physical semi-relativistic solution $U_{\alpha 3}$ where the singular potentials [1] with the singularity ρ^{-3} can be included because the last term in (25.8) of semi-relativistic equation (25.6) has the asymptote ρ^{-4} at the origin. Using solutions $U_{\alpha 3}$, we can find the expectation values for all terms of the realistic Hamada–Johnston potential [1] which has singularity ρ^{-6}.

Now we can see that for large-scale distances F, the semi-relativistic equation (25.3) reduces into the Schrödinger equation. For the nucleons localized around the center of force of about 1 fm, we have $C_F = 0.011$. In the region of the repulsive core for nucleons' interaction potential or typical distances for the quark–quark interaction region (0.4 fm), we have $C_F = 0.07$. For the electrons in the first Bohr orbit, $C_F = 1.3 \cdot 10^{-5}$. These results show that in the theory of atomic spectroscopy, we can calculate the relativistic corrections with sufficient accuracy in the first approximation of the perturbation theory A[15]. However, for calculations of the nuclear energy levels or consideration of the excited bound quark systems, we must include the higher order perturbations, and all solutions

$U_{\alpha 1}$, $U_{\alpha 3}$ and $U_{\alpha 4}$ of the semi-relativistic equation (25.3) must be used. The solutions $U_{\alpha 1}$ and $U_{\alpha 2}$ for the model harmonic oscillator potential were found in Ref. A[15].

25.3 The Approximate System of Integral Semi-relativistic Equations

General solutions of the Schrödinger equation in the potential representation can be presented like the system of integral equations A[19]. In our calculations, we will use the approximate method A[15], where one-particle wave functions are expressed as a product of the unperturbed solution and the function which depends on the perturbation potential.

Using the multiplicative perturbation theory considered in papers A[1, 4], we will consider that the solutions of the Schrödinger equation must essentially differ from the solutions of the semi-relativistic equation (25.3) only at the origin (25.6). For consideration of bound system nucleons or quarks, the model harmonic oscillator potential $V_1(r) = m\omega^2 r^2/2$ can be used A[15].

The radial wave functions U_{nL1} and singular non-physical solutions F_{nL} of the Schrödinger equation A[15] and approximate solutions of semi-relativistic equation for the this model potential can be presented in the following way:

$$U_{nLi} = e^{-0.5\rho^2} \rho^{\beta(i)} \sum_{k=0}^{n-1} a_k \rho^{2k} \quad \text{and} \quad F_{nL} = e^{-0.5\rho^2} \rho^{-L} \sum_{k=0}^{\infty} b_k \rho^{2k},$$

where

$$\rho^2 = \frac{r^2}{F^2}, \quad F = \sqrt{\frac{\hbar}{m\omega}}, \quad n = 1, 2, 3, \ldots, \quad i = 1, 2, 3, 4,$$

$$\beta(1) = L + 1, \quad L = 0, 1, 2, \ldots,$$

$$a_0 = 1, \quad a_{k+1} = \frac{k - 0.5(\varepsilon_{nL,i} - \beta(1) - 0.5)}{(k+1)(k + \beta(1) + 0.5)} a_k,$$

$$b_0 = 1, \quad b_{k+1} = \frac{k - 0.5(\varepsilon_{nL,i} - L - 0.5)}{(k+1)(k - L + 0.5)} b_k.$$

(25.8)

Approximate solutions U_{nL3} and U_{nL4} of the semi-relativistic equation (25.3) must coincide with U_{nL1} for $\rho > 1$. Solutions U_{nL3} and U_{nL4} differ from U_{nL1} only at distances $\rho < 1$, and in this region, we obtain approximate solutions U_{nL3} and U_{nL4} by substituting in the last formula $\beta(3) = L+3$, $L = 0, 1, 2, \ldots$, $\beta(4) = -L+2$, $L \le 1$.

The eigenvalues for the wave function U_{nL1} and zero approximation of eigenvalues for U_{nL3} and U_{nL4} have the same term

$$E_{nL1} = \varepsilon_{n11}\hbar\omega \quad \varepsilon_{nL1} = 2n + \beta(1) - \frac{3}{2}, \quad n = 1, 2, 3, \ldots. \quad (25.9)$$

The partial solutions (25.3) must satisfy the following boundary conditions at the origin:

$$\lim_{\rho \to 0} U_{\alpha1} \cdot \rho^{-L-1} = 1, \quad \lim_{\rho \to 0} U_{\alpha2} \cdot \rho^{L} = 1,$$
$$\lim_{\rho \to 0} U_{\alpha3} \cdot \rho^{-L-3} = 1, \quad \lim_{\rho \to 0} U_{\alpha4} \cdot \rho^{L-2} = 1 \quad (25.10)$$

and at infinity

$$\lim_{\rho \to \infty} \varphi_{\alpha1} \cdot \rho^{L+1} = 0, \quad \lim_{\rho \to \infty} \varphi_{\alpha2} \cdot \rho^{-L} = \infty,$$
$$\lim_{\rho \to \infty} \varphi_{\alpha3} \cdot \rho^{L+3} = 0, \quad \lim_{\rho \to \infty} \varphi_{\alpha4} \cdot \rho^{-L+2} = 0. \quad (25.11)$$

Using the method of indefinite coefficients [3] for the solutions U_{nLi} and F_{nL}, we can obtain integral equations for bound states

$$\varphi_{1i}U_{nLi} = U_{nLi} + \frac{U_{nLi}}{W_{0i}} \int_0^\rho F_{nL}CV_\delta\varphi_1 U_{nLi}d\rho_1$$
$$- \frac{F_{nL}}{W_{0i}} \int_0^\rho U_{nLi}CV_{\delta i}\varphi_1 U_{nLi}d\rho_1, \quad (25.12)$$

$$\Delta E_{nLi} = \frac{\int_0^\infty U_{nLi}V_{Di}\varphi_{li}U_{nLi}d\rho}{\int_0^\infty U_{nLi}\varphi_{li}U_{nLi}d\rho}, \quad V_{\delta i} = V_D(\rho) - \Delta E_{nLi},$$
$$V_D = \frac{C_1}{CF^4}D(\rho), \quad i = 1, 3, 4, \quad (25.13)$$

where W_{0i} are the Wronskian of the wave functions U_{nLi} and the linearly independent solution F_{nL} for model harmonic oscillator potential

$$W_{0i} = (2L_i + 1)\left(\frac{m\omega}{\hbar}\right)^{1/2}, \quad L_i = L + \beta(i) - \beta(1). \tag{25.14}$$

Practically, we used the approach similar to that used in the paper A[15] for the solution U_{nL1}, but here additional semi-relativistic solutions U_{nL3} and U_{nL4} were included. Solving integral equations (25.12) and (25.13) by the iteration method A[15], we obtain three solutions U_{nL1}, U_{nL3} and U_{nL4} and three different energies

$$E_{nLi} = E_{nL1} + \Delta E_{nLi}, \quad i = 1, 3, 4. \tag{25.15}$$

The obtained integral–differential equations (25.12) and (25.13) can also be used for finding wave functions and energies of bounded quark systems in the simplified semi-relativistic approach which is less complicated than the exact method A[19].

The obtained mathematical results about solutions A[19] of semi-relativistic equation (25.3) represent interesting physical consequence. For example [2], the very heavy baryons which consist of three charmed or b quarks of different colors and belong to the lowest multiplet $I = 3/2$ for $n = 1$, $L = 0$, according to semi-relativistic model, can be presented like the anti-symmetrical state of three identical quarks in three different semi-relativistic states $U_{\alpha 1}$, $U_{\alpha 3}$ and $U_{\alpha 4}$ with different energies.

25.4 Semi-relativistic Model for Charmonium and Bottomonium

For theoretical investigation of quark interactions, it is important to study the energy levels of charmonium and bottomonium. Usually, a simple potential model can be applied for this. This potential consists of long-range linear potential and a short-range Coulomb potential, generated by exchange of a single-color gluon between the quarks [2]:

$$V(r) = ar + c - \frac{4}{3}\frac{\alpha_s(m_Q)}{r}. \tag{25.16}$$

Also, additionally, spin–spin, tensor interactions and relativistic corrections to the mass are included [4] like the first-order corrections for the Schrödinger solutions. The relativistic corrections for the mass are very important for nuclear spectra calculations A[15, 16]. We can suppose that these corrections must be significant for the heavy-quark bound states. Expanding kinetic energy in the power series, we obtain

$$E - m_0 c^2 = \frac{p^2}{2m_0} - \frac{p^2}{2m_0} R + \frac{p^2}{2m_0} R^2 -,$$

$$R = \frac{1}{2m_0 c^2} \frac{p^2}{2m_0}.$$

(25.17)

Relativistic kinetic energy depends on the relation between kinetic energy with rest energy of the quarks and antiquarks. In the semi-relativistic approach, the second term which represents the relativistic correction for the kinetic energy or mass in the presented expansion was included in the Hamiltonian like the operator. This approach is sufficiently exact only when $R \ll 1$. Using the Hamiltonian operator with correction operator for kinetic energy (25.1), we obtain the fourth-order differential equation (25.3), which we practically solved by transforming to the integral equation (25.12). This equation was used for calculations of one-nucleon energy levels for heavy nuclei A[15, 16]. Here, nucleons are moving in the region of about 10 fm when the parameter $C_F = 10^{-6}$ is sufficiently small. In this case, the terms (25.1) and (25.2) can be included only like perturbations. We have a more complicated situation in the quark–antiquark systems. In this case, the solutions U_{nL3}, U_{nL4} must be included.

For $L = 0$, we have three different semi-relativistic solutions $U_{\alpha 1, L+1}$, $U_{\alpha 3, L+3}$ and $U_{\alpha 2, -L+2}$, which at the origin have the different asymptotic behavior r^{L+1}, r^{L+3} and r^{-L+2} and different energies (25.14). Inversely, we can say that three semi-relativistic states transform to one for the non-relativistic state U_{nL} in a non-relativistic region.

Having the aim to investigate the properties of semi-relativistic dynamics and their connections with relativistic corrections for mass or kinetic energy in the heavy quark–antiquark systems $c\bar{c}$ and $b\bar{b}$, we used the model harmonic oscillator potential

$$V_{1q}(r) = V_{0q} - \frac{m_q \omega_q^2 r^2}{2}, \qquad (25.18)$$

for which the wave functions (25.8) U_{nL1} and energies (25.9) E_{nL1} for $\beta(1) = L+1$ are known. In this case, the integral equation (25.12) can be solved exactly, and relativistic corrections for mass and potential can be calculated like perturbations. In the first-order approximation, these corrections in the heavy-quark $c\bar{c}$ and $b\bar{b}$ systems were calculated in Ref. [5], but this approach is sufficient only for atomic nuclei A[14].

We can see (25.10) that relativistic corrections for the mass and potential are proportional to the constant C_F. These constants for charmonium C_{Fc} and bottomonium C_{Fb} with masses [6]

$$m_c = 1.55\,\text{GeV}, \quad m_b = 4.88\,\text{GeV} \qquad (25.19)$$

for $F = 10^{-15}$ m are

$$C_{Fc} = 4,05 \cdot 10^{-3}, \quad C_{Fb} = 4.09 \cdot 10^{-4} \qquad (25.20)$$

greater than those for the heavy nucleus $C_F = 10^{-6}$. In this case, the relativistic corrections for quark masses and interaction potentials in hadrons must also be significantly greater. The potentials (25.16) are usually defined in the region $0.2\,\text{fm} \le r \le 0.8\,\text{fm}$. At the point $r_0 = 0.2\,\text{fm}$ for charmonium, we have $C_{r_0c} = 0.101$, and more exact equations must be used A[19] than our approach where relativistic corrections are included in the kernel of the integral equation (25.12) like perturbation.

Parameters $\hbar\omega = 0.2985\,\text{GeV}$ and $V_{0b} = -0.7553\,\text{GeV}$ were obtained for the model harmonic oscillator potential (25.21) from the bound states $1S$ and $2S$ of $b\bar{b}$ presented in Ref. [7]. Then, solving the integral equation (25.12) by the iteration method A[15], from (25.13) and (25.15), we obtain

$$E_{nL} = E_{\text{OSC}} + \Delta E_{nL}, \quad E_{\text{OSC}} = -V_0 + \left(2n + L + 1 - \frac{3}{2}\right) \cdot \hbar\omega,$$

$$\Delta E_{nL} = \Delta E_{nL,m} + \Delta E_{nL,p}.$$

$$(25.21)$$

Table 25.1.   The mass spectra M_{nL} of $b\bar{b}$ bound states and relativistic corrections.

nL	State	M_{nL}(GeV)	$\Delta E_{nL,m}$(GeV)	$\Delta E_{nL,p}$(GeV)
1S	Y(9.460)	9.4600	−0.01066	0.01870
2S	Y′(10.0233)	10.0200	−0.04474	0.01527
3S	Y″(10.3553)	10.441	−0.21843	0.01372
4S	Y‴(10.580)	10.840	−0.41176	0.01375
1P	χ_b(9.9002)	9.7816	0.01547	0.01564

Here, E_{OSC} are the energies of the eigenstates of the Schrödinger equation for the harmonic oscillator potential (25.18) with the above-presented parameters. $\Delta E_{nL,m}$ and $\Delta E_{nL,p}$ are relativistic corrections for mass and potential, consequently. The mass M_{nL} for $b\bar{b}$ bound states are related to the bound energies

$$M_{nL} = 2m_q + E_{nL}, \qquad (25.22)$$

where m_q is the mass of quark q expressed in electron volts. The experimental masses of the presented mesons, obtained mass spectra m_{nL} of $b\bar{b}$ bound states [6, 7] and the relativistic corrections for mass $\Delta E_{nL,m}$ and potential energy $\Delta E_{nL,p}$ are presented in Table 25.1.

From the calculated results, we see that for $b\bar{b}$ state, 1S relativistic corrections are small. With relativistic correction to mass for 3S state, we obtained sufficiently good coincidence between M_{nL} and the experimental state Y″(10.3553).

We have another situation for the charmonium states [7]. For this reason, the relativistic corrections obtained for the mass were larger, and we provided integration of the integral equation (25.12) in the interval $0.28\,\text{fm} \leq r \leq 1\,\text{fm}$. For better coincidence with the experiment, the third term in expansion of (25.17) must be included like perturbation with a different sign and the expectation values of the second and the third terms can partially compensate each other. Taking different $V_{0c} = -0.4564\,\text{GeV}$, the same $\hbar\omega$ and substituting E_{nL} by E_{OSC} in (25.13), (25.15) and (25.22), we present the experimental masses of the presented mesons, mass spectra M_{nL} of $c\bar{c}$ bound states and the relativistic corrections for mass $\Delta E_{nL,m}$ and potential energy $\Delta E_{nL,p}$ in Table 25.2.

Table 25.2. The mass spectra M_{nL} of $c\bar{c}$ bound states and relativistic corrections.

nL	State	M_{nL}(GeV)	$\Delta E_{nL,m}$(GeV)	$\Delta E_{nL,p}$(GeV)
1S	$J/\psi(3.097)$	3.097	-0.06913	0.04594
2S	$\psi'(3.686)$	3.528	-0.4582	0.04867
3S	$\psi''(4.040)$	3.736	-0.6001	0.04691
1P	$\chi_c(3.500)$	3.504	0.06886	0.04543

Great relativistic corrections for mass of $c\bar{c}$ bound states show that the additional term H_m in the Hamiltonian (25.16) represents only minimal relativity and are presented in Table 25.2. Considering that the third term in the kinetic energy expansion (25.17) can partially compensate the term H_m, in our case we obtained sufficiently good coincidence of the calculated $c\bar{c}$ mass spectra with the experiment. It is important to note that relativistic corrections to mass and potential in 1S state are small.

Using approximate functions $U_{\alpha3}$ and $U_{\alpha4}$ instead of U_{nL1} in (25.12) for quantum numbers $n = 1$ and $L = 0$, we obtained two additional semi-relativistic eigenvalues (25.15). For semi-relativistic eigenfunctions $U_{\alpha3}$ and $U_{\alpha4}$ for states $n = 1$ and $L = 0$, we obtained the following eigenvalues: $E_{n,L+3} = -0.3098$ GeV, $E_{n,-L+2} = -0.3065$ GeV for bottomonium and $E_{n,L+3} = -0.001176$ GeV, $E_{n,-L+2} = -0.007334$ GeV for charmonium. Also, we obtained the semi-relativistic eigenvalues E_{nL} for bottomonium as -0.3000 GeV and for charmonium as -0.002998 GeV in the state $1S$ for the solution $U_{\alpha1}$ which possess the same asymptotic at the origin like the Schrödinger solution U_{nL}. Other non-relativistic states nL for $L = 1$ also split into three different semi-relativistic states with different energies.

25.5 Conclusions

In this chapter, we presented three different sets of physical solutions of semi-relativistic equations with different asymptotes at the origin. For $L = 0$, we have three different semi-relativistic solutions $U_{\alpha1,L+1}$, $U_{\alpha3,L+3}$ and $U_{\alpha2,-L+2}$ with different asymptotic behaviors r^1, r^3 and

r^2 near the origin and different energies. These three semi-relativistic states in the non-relativistic region transform to one non-relativistic state U_{nL}. The dependence of relativistic corrections for mass to one-nucleon levels of energies in the harmonic oscillator well on the asymptote of radial wave functions at the origin was found in the paper A[20].

For example, theoretically predicted baryons [2] $\Omega(ccc)$ and $\Omega(bbb)$ can consist of three identical quarks in the same non-relativistic state $n = 1$ and $L = 0$ only when colors were introduced. Using our approach, we obtained three different semi-relativistic states. In the semi-relativistic approach, we have no contradictions with the Pauli exclusion principle for presented baryons and for the existence of exotic mesons [8] $f_0(600)$, $K(800)$, $a_0(980)$ and $f_0(980)$, which consist of the following quark–antiquark pairs $ud\bar{u}\bar{d}$, $ud\bar{s}\bar{d}$, $ds\bar{u}\bar{s}$ and $us\bar{u}\bar{s}$ having quantum numbers that are impossible for the standard quark models. The spins of these mesons' semi-relativistic solutions $U_{\alpha 1,L+1}$, $U_{\alpha 3,L+3}$ and $U_{\alpha 2,-L+2}$ with the different asymptotic behavior r^1, r^3 and r^2 can be explained without any colors using semi-relativistic quantum states $U_{\alpha 1,L+1}$, $U_{\alpha 3,L+3}$ and $U_{\alpha 2,-L+2}$ with the different asymptotes r^1, r^3 and r^2 separating particles and antiparticles. Also, a semi-relativistic model can show that the dark matter consists of the second quarks and antiquarks combination $u, u, d, \bar{u}, \bar{u}, \bar{d}$ using semi-relativistic solutions $U_{\alpha 1,L+1}$, $U_{\alpha 3,L+3}$ and $U_{\alpha 2,-L+}$ with the different asymptotic behaviors r^1, r^3 and r^2, which represent potential barriers separating quarks and antiquarks. Because non-relativistic states U_{nL}, when relativistic corrections are included, split in three different semi-relativistic states A[19] $U_{\alpha 1,L+1}$, $U_{\alpha 3,L+3}$ and $U_{\alpha 2,-L+2}$, this conflict is removed A[20]. The split of these bound states is small and can be evaluated using the semi-relativistic approach presented.

The semi-relativistic model in the single-particle approach was considered and the integral equations for the singular realistic nucleon–nucleon and quark–antiquark potentials were obtained. The significant relativistic corrections for mass of charmonium and bottomonium states for the model harmonic oscillator potential using integral equations were included. The calculated masses for excited states in our approach are in sufficient good coincidence with the experiment. However, our minimal relativity is not sufficient for energy evaluation of excited states of charmonium.

References

1. Heyde, K. L. G. (1994). *The Nuclear Shell Model*, Springer-Verlag, Berlin.
2. Griffiths, D. (2008). *Introduction to Elementary Particles*, *Wiley-VCH*, *Veinheim*, Germany, p. 454.
3. Gutter, R. S. and Janpolski, A. R. (1976). *Differential Equations*, Vyshaya Shkola, Moscow (in Russian).
4. Kang, J. S. (1979). Heavy-quark bound states and a potential model with relativistic corrections, *Phys. Rev. D* **20**, 29781999.
5. Hey, A. J. G. and Kelly, R. L. (1983). Quark model, *Phys. Rep.* **96**, 71–80.
6. Ebert, D., Faustov, R. N. and Galkin, V. O. (2000). Quark-antiquark potential with retardation and irradiative contribution on the heavy quarkonium spectra, *Phys. Rev. D* **62**, 034014-1, 1–10.
7. Liu, B., Shen, P. N. and Chiang, H. C. (1997). Heavy quarkonium spectra and dissociation in hot and dense matter, *Phys. Rev. C*, **55**, 3021–3026.
8. Kisak, P. F. (2016). *The Meson, A Hadronic Subatomic Particle*, Lexington, KY, USA, p. 57.

Bibliography of the Authors

The papers cited in the book: A. J. Janavičius and D. Jurgaitis, *Quantum Mechanics in Potential Representation and Applications.*

1. Janawiczius, A.J. and Kwiatkowski, K. (1978) *Schrödinger Equation in Potential Representation for Saxon-Woods Potential Representation for the Case of S-Wave.* Report No, 1001/PL, Krakow, p. 12.
2. Janawiczius, A.J. and Planeta, R. (1978) *A General Solution of the Schrödinger in the Potential Representation.* Report No. 1018/PL, Krakow, p. 13.
3. Janavičius, A.J. (1975) *"Using Theory of Complex Orbital Momentum for Finding Constants of Model Potential from Experimental Phase Shifts,"* in *Lithuanian Physical Collection*, 15, No. 3, p. 369 (in Russian).
4. Janavičius, A.J. (1998) *"The General Solution of the Schrödinger Equation for Positive Energies and Bound States in the One-Particle Case,"* in *Lithuanian Journal of Physics*, 38, No. 5, pp. 437–441.
5. Janavičius, A.J. and Jurgaitis, D. (1985) *"Schrödinger Integral Equation in the Potential Representation for the Case of Positive Energy,"* in *Lithuanian Physical Collection*, XXV, No. 3, pp. 34–38.

6. Janavičius, A.J. and Jurgaitis, D. (1986) *"Separation of the Scattering Matrix with Short Range Potential from the Background Coulomb Field,"* in *Lithuanian Physical Collection*, XXVI, No. 3, pp. 273–278.

7. Janavičius, A.J. and Žilinskas, K. (2013) *"The General Solution of the Schrödinger Equation for Bound States,* in *Canadian Journal of Physics*, 91, pp. 378–381.

8. Janavičius, A.J. (2000) *"The Method of Potential Representation for Nonspherical Potential,"* *Lithuanian Journal of Physics*, 40, No. 4, pp. 266–270.

9. Janavičius, A.J. (1998) *"Green's Functions and Nonphysical Solutions,"* in *Lithuanian Journal of Physics*, 38, No. 5, pp. 431–435.

10. Janavičius, A.J., Jurgaitis, D. and Turskienė, S. (2011) *"Potential Representation Method for the Equation,"* in *Mathematical Modelling and Analysis*. Taylor & Francis and VGTU, Vol. 16, No. 3, September, pp. 442–450.

11. Janavičius, A.J. (1982) *"Solution of Schrödinger Equation in Potential Representation for Coulomb Interactions,"* in *Lithuanian Physical Collection*, XXII, No. 2, pp. 20–28.

12. Baškienė and Janavičius, A.J. (2004) *"Transformations of the Hamiltonian For Jastrow's Correlation Method,"* *Šiauliai University Mathematical Seminar*, 7, pp. 5–11.

13. Janavičius, A.J. (1983) *"Relativistic Corrections for One-Neutron Levels in Square Well of Harmonic Oscillator Potential,"* *Information of Higher Education Institutions*, 12, pp. 14–17 (in Russian).

14. Janavičius, A.J. and Bakštys, A. (1993) *"Relativistic Corrections to the One-Nucleon Energy Levels of ^{208}Pb,"* in *Acta Physica Polonica B*, No. 12, Vol. 24, pp. 1981–1987.

15. Janavičius, A.J. (1996) *"Relativistic Corrections in the Average Field of Neutron and Proton Shells,"* *Acta Physica Polonica B*, 27, pp. 2195–2205.

16. Janavičius, A.J. (2001) *"Shell Stability of Heaviest Atomic Nuclei in the Semi-Relativistic Model,"* *Lithuanian Journal of Physics*, 41, No. 3, pp. 232–236.

17. Janavičius, A.J. (2006) *"Mathematical Methods in the Semi-Relativistic Single-Particle Model for Superheavy Nuclei,"* in *The 10th World Multi-Conference on Systemics, Cybernetics and Informatics.* Orlando, Florida, July 16–19, pp. 66–68.

18. Janavičius, A.J. and Bagdonaitė, R. (2001) *"Semi-Relativistic Nuclear Shell Model,"* Lithuanian Journal of Physics, 41, No. 3, pp. 150–153.

19. Janavičius, A.J., Jurgaitis, D. and Korsakienė, D. (2004) *"Integral–Differential Equations for Semi-Relativistic Nuclear Shell Model,"* Acta Physica Polonica B, 35, pp. 757–765.

20. Janavičius, A.J. (2005) *Relativistic Corrections of Charmed and Bottom Mesons, Proceedings of Scientific Seminar of the Faculty of Physics and Mathematics,* Šiauliai University, **8**, pp. 35–45.

21. Janavičius, A.J. (2018) *"Particles and Quantum Waves Diffusion in Physical vacuum,"* Results in Physics, 117, pp. 148–151.

22. Janavičius, A.J. (2012) *"Quantum Diffusion of Electron in Hydrogen Atom,"* in *17 International Conference on Mathematical Modelling and Analysis. Abstracts,* June 6–9, Sigulda, Latvia, p. 55.

23. Janavičius, A.J. and Purlys, R. (2012) "Investigation of Superdiffusion of Metastable Vacancies by Bragg Diffraction of X-Rays," in *4th International Conference Radiation Interaction with Material and Its Use in Technologies 2012. Program and Materials,* Kaunas, Lithuania, May 14–17, pp. 612–615.

24. Janavičius, A.J., Jurgaitis, D. and Kiriliauskaitė, K. (2011) *"The Potential Representation of the Schrodinger Equation for Bound States in Many Particles Case,"* in *16 International Conference on Mathematical Modelling and Analysis. Abstracts,* May 25–28, Sigulda, Latvia, p. 61.

25. Janavičius, A.J. and Žilinskas, K. (2011) *"The General Solution of the Schrödinger Equation for Bound States,"* in *16 International Conference on Mathematical Modelling and Analysis. Abstracts,* May 25–28, Sigulda, Latvia, p. 62.

26. Janavičius, A.J. and Jurgaitis, D. (2012) *"Superposition Energy for Bound States of Two Potentials in the Schrodinger Equation,"* in *17 International Conference on Mathematical*

Modelling and Analysis. Abstracts, June 6–9, Sigulda, Latvia, p. 61.

27. Purlys, R., Janavičius A.J. *et al.* (2016) *Method of Creation of Defects Using X-Ray Radiation and Electric Field and Its Application, United States Patent* No: US 9,530,650 B2, Date of the Patent: Dec. 27.

Index

CPSIA information can be obtained
at www.ICGtesting.com
Printed in the USA
BVHW040510240720
584450BV00006B/19